McDougal Littell

# Algebra 1
## Concepts and Skills

Larson    Boswell    Kanold    Stiff

# Standardized Test Practice Workbook
# Teacher's Edition

The Standardized Test Practice Workbook provides
practice exercises for every lesson in a standardized
test format. Included are multiple-choice, quantitative-
comparison, and multi-step problems. The Teacher's
Edition includes the student workbook and the answers.

**McDougal Littell**
A HOUGHTON MIFFLIN COMPANY
Evanston, Illinois • Boston • Dallas

ISBN: 0-618-07868-1

23456789-DWI-04 03 02 01

# Contents

NAME _____ DATE _____

# Standardized Test Practice

**For use with pages 3–8**

**TEST TAKING STRATEGY   Spend no more than a few minutes on each question.**

1. **Multiple Choice**   What is the value of the expression $7 - x$ when $x = 2$?

   Ⓐ  14          Ⓑ  $-5$          Ⓒ  $-9$

   Ⓓ  5          Ⓔ  $-19$

2. **Multiple Choice**   What is the value of the expression $\frac{24}{x}$ when $x = 3$?

   Ⓐ  27          Ⓑ  21          Ⓒ  8

   Ⓓ  3          Ⓔ  $\frac{1}{8}$

3. **Multiple Choice**   Evaluate $3y - 6$ when $y = 4$.

   Ⓐ  6          Ⓑ  $-18$          Ⓒ  18

   Ⓓ  $-6$          Ⓔ  $-12$

4. **Multiple Choice**   Find the area of a triangle if the height is 6 inches and the base is 9 inches. (Hint: Area of a triangle is $A = \frac{1}{2}bh$.)

   Ⓐ  7.5          Ⓑ  18          Ⓒ  15

   Ⓓ  54          Ⓔ  27

5. **Multiple Choice**   Evaluate $\frac{6}{7}x$ when $x = \frac{2}{3}$.

   Ⓐ  $\frac{8}{10}$          Ⓑ  $1\frac{11}{21}$          Ⓒ  $\frac{12}{7}$

   Ⓓ  $\frac{4}{5}$          Ⓔ  $\frac{4}{7}$

6. **Multiple Choice**   The variable expression for 3 times $x$ plus 1.2 is __?__ .

   Ⓐ  $1.2x + 3$     Ⓑ  $3x + 3.6$

   Ⓒ  $3x + 1.2$     Ⓓ  $x + 1.2 \cdot 3$

   Ⓔ  $\frac{x}{3} + 1.2$

7. **Multiple Choice**   Which is the variable expression for 26 divided by $w$?

   Ⓐ  $\frac{26}{w}$          Ⓑ  $\frac{w}{26}$

   Ⓒ  $26w$          Ⓓ  $26 + w$

   Ⓔ  $26 - w$

8. **Multiple Choice**   A bat can eat 0.05 pounds of mosquitoes in an hour. If it eats for 6.5 hours, how many pounds of mosquitoes will it consume?

   Ⓐ  0.6 lb          Ⓑ  0.325 lb

   Ⓒ  0.13 lb          Ⓓ  3.25 lb          Ⓔ  1.3 lb

9. **Multiple Choice**   How many miles do you travel if you bike for 2 hours at an average speed of 6 miles per hour?

   Ⓐ  3 mi          Ⓑ  12 mi          Ⓒ  8 mi

   Ⓓ  4 mi          Ⓔ  $\frac{1}{3}$ mi

**Quantitative Comparison**   In Exercises 10–12, choose the statement below that is true about the given numbers.

   Ⓐ  The number in column A is greater.

   Ⓑ  The number in column B is greater.

   Ⓒ  The two numbers are equal.

   Ⓓ  The relationship cannot be determined from the given information.

|  | Column A | Column B |
|---|---|---|
| 10. | $\frac{6}{x}$ when $x = 4$ | $3y$ when $y = 3$ |
| 11. | The perimeter of a square with 1.5 m sides. | The perimeter of a triangle with 2 m sides. |
| 12. | $3z$ when $z = 2$ | $\frac{1}{2}z$ when $z = 10$ |

NAME _____ DATE _____

# *Standardized Test Practice*

For use with pages 9–14

**TEST TAKING STRATEGY**   **As soon as the testing time begins, start working. Keep a steady pace and stay focused on the test.**

1. *Multiple Choice*   Which algebraic expression is equivalent to $6 \cdot 6 \cdot 6 \cdot 6 \cdot 6$?

   (A) $6(5)$      (B) $5^6$      (C) $7776$

   (D) $30$      (E) $6^5$

2. *Multiple Choice*   Express the meaning of the power $3^7$ in numbers.

   (A) $7 \cdot 7 \cdot 7$   (B) $3 \cdot 3 \cdot 3 \cdot 3 \cdot 3 \cdot 3 \cdot 3$

   (C) $3(7)$      (D) $2191$

   (E) $3 \cdot 3 \cdot 3 \cdot 3 \cdot 3 \cdot 3$

3. *Multiple Choice*   Evaluate the expression $x^5$ when $x = 3$.

   (A) $125$      (B) $243$      (C) $15$

   (D) $8$      (E) $81$

4. *Multiple Choice*   Evaluate the expression $(x^2) - y$ when $x = 5$ and $y = 3$.

   (A) $22$      (B) $7$      (C) $4$

   (D) $49$      (E) $28$

5. *Multiple Choice*   Evaluate the expression $(m + p)^3$ when $m = 4$ and $p = 3$.

   (A) $67$      (B) $31$      (C) $21$

   (D) $1728$      (E) $343$

6. *Multiple Choice*   Evaluate the expression $2x + (y^2)$ when $x = 7$ and $y = 4$.

   (A) $25$      (B) $17$      (C) $144$

   (D) $30$      (E) $22$

7. *Multiple Choice*   Evaluate the expression $(3a)^4$ when $a = 2$.

   (A) $625$      (B) $162$      (C) $48$

   (D) $1296$      (E) $24$

8. *Multiple Choice*   Halie's bedroom is a cube measuring 10 feet along the base of each wall and 10 feet high. Which expression is used to calculate the square footage needed to wallpaper her room?

   (A) $10^4$      (B) $10^3$

   (C) $4(10^2)$      (D) $4^2 \cdot 10$

   (E) $(4 \cdot 10)^2$

9. *Multiple Choice*   A cylindrical vase is 14 inches high and has a radius of 5 inches. Approximate the volume in cubic inches using the formula 3.14 times the square of the radius times height.

   (A) $3077$      (B) $353.14$      (C) $1099$

   (D) $439.6$      (E) $1380$

10. *Multi-Step Problem*   You have a large rectangular fish tank measuring 3 feet across by 2 feet deep by 2 feet high. You want to fill it $\frac{3}{4}$ of the way with water.

   a. What is the volume of the fish tank in cubic feet?

   b. If one cubic foot is approximately equal to 7.48 gallons, how many gallons of water can the tank hold?

   c. How many gallons will fill the tank $\frac{3}{4}$ full?

   d. Would a fish tank measuring 3 feet across by 4 feet deep by 2 feet high double the volume of your current fish tank?

NAME _____ DATE _____

# *Standardized Test Practice*

**For use with pages 15–21**

**TEST TAKING STRATEGY** **If you can, check an answer using a method that is different from the one you used originally, to avoid making the same mistake twice.**

1. *Multiple Choice* Evaluate the expression $4x^2 - 2$ when $x = 3$.

   (A) 47      (B) 142      (C) 22

   (D) 72      (E) 34

2. *Multiple Choice* Evaluate the expression $126 \div (2y^3)$ when $y = 3$.

   (A) $\frac{7}{12}$      (B) 1701      (C) 7

   (D) $2\frac{1}{3}$      (E) 567

3. *Multiple Choice* Evaluate the expression $\frac{88}{x^4} + 6x$ when $x = 2$.

   (A) 23      (B) 3.14      (C) 17.5

   (D) 4.4      (E) 13.5

4. *Multiple Choice* Evaluate the expression $\frac{2}{3}x^2 - 4$ when $x = 9$.

   (A) 50      (B) 77      (C) 2

   (D) 117.5      (E) 32

5. *Multiple Choice* Evaluate the expression $6 \cdot (3^2 + 6)$.

   (A) 72      (B) 66      (C) 60

   (D) 36      (E) 90

6. *Multiple Choice* Insert grouping symbols into $1 + 4 \cdot 3^2 + 2$ to produce the value 45.

   (A) $1 + (4 \cdot 3)^2 + 2$    (B) $1 + 4 \cdot (3^2 + 2)$

   (C) $1 + 4 \cdot (3^2) + 2$    (D) $(1 + 4) \cdot 3^2 + 2$

   (E) $(1 + 4 \cdot 3)^2 + 2$

7. *Multiple Choice* Evaluate $\frac{1}{3}(48 + 3) - 2^3$.

   (A) 41      (B) 9      (C) 24

   (D) 40      (E) 11

8. *Multiple Choice* What is the value of $\frac{15 - 6 \cdot 2}{8 \cdot (4^2 - 7)}$?

   (A) $2\frac{1}{4}$      (B) $\frac{1}{24}$      (C) $\frac{3}{8}$

   (D) $\frac{3}{121}$      (E) $\frac{3}{4}$

9. *Multiple Choice* If your state charges a 6% sales tax on all items except food and clothing, what is the sales tax if you bought groceries for $22.15, a pair of jeans for $32.00, a video for $16.00, and books for $35.00?

   (A) $6.31      (B) $4.39      (C) $4.98

   (D) $2.10      (E) $3.06

*Quantitative Comparison* In Exercises 10–12, choose the statement below that is true about the given numbers.

   (A) The number in column A is greater.

   (B) The number in column B is greater.

   (C) The two numbers are equal.

   (D) The relationship cannot be determined from the given information.

|     | Column A | Column B |
| --- | --- | --- |
| 10. | $(19 - 4)^2 + 5$ | $19 - 4^2 + 5$ |
| 11. | $18 + 9 \div 3 - 6$ | $15 \div 3 \cdot 2^2 - 5$ |
| 12. | $3 + 6^2 \div 4 + 2$ | $12 \div 3 \cdot 2^2 - 1$ |

# *Standardized Test Practice*

**For use with pages 24–29**

**TEST TAKING STRATEGY    Avoid spending too much time on one question.**

1. *Multiple Choice*    The solution of the equation $7x + 2 = 23$ is __?__ .

   (A)  4    (B)  3    (C)  14    (D)  12    (E)  5

2. *Multiple Choice*    The solution of the equation $15 - 2y = 7$ is __?__ .

   (A)  5    (B)  4    (C)  2    (D)  3    (E)  6

3. *Multiple Choice*    The solution of the equation $10x - 6 = 10 + 6x$ is __?__ .

   (A)  1    (B)  4    (C)  3    (D)  5    (E)  2

4. *Multiple Choice*    Determine which equation has a solution of $x = 3$.

   (A)  $4x + 6 = 18$    (B)  $16 = 12x + 1$

   (C)  $18 - 3x = 12$    (D)  $4 + 2x = 3x - 1$

   (E)  $5 + 2x = 13$

5. *Multiple Choice*    Determine which equation has a solution of $a = 5$.

   (A)  $2a + 6 = 10$    (B)  $32 = 7a + 4$

   (C)  $12 - a = 4$    (D)  $25 - 3a = 10$

   (E)  $16 - a^2 = 8$

6. *Multiple Choice*    Which equation does not have a solution of $w = 6$?

   (A)  $3w - 4 = 14$    (B)  $10w - 22 = 38$

   (C)  $12 + 2w = 18$    (D)  $w^2 + 3 = 39$

   (E)  $16 = 22 - w$

7. *Multiple Choice*    Which equation is a translation of "3 times a number decreased by 6 gives 18"?

   (A)  $3x \div 6 = 18$    (B)  $6 - 3x = 18$

   (C)  $6 - 3x = 18$    (D)  $3x + 6 = 18$

   (E)  $3x - 6 = 18$

8. *Multiple Choice*    The perimeter of a square is 16 cm. Which equation models the situation?

   (A)  $4x = 16$    (B)  $\frac{1}{4}x = 16$

   (C)  $x^2 = 16$    (D)  $x + x + x = 16$

   (E)  $(4x)^2 = 16$

9. *Multiple Choice*    For which of the following is $x = 3$ a solution?

   (A)  only $2x < 8$

   (B)  only $10 - 2x < 3$

   (C)  only $15 > 4x - 6$

   (D)  both $2x < 8$ and $10 - 2x < 3$

   (E)  both $2x < 8$ and $15 > 4x - 25$

10. *Multiple Choice*    Match the sentence "the sum of 10 and twice $x$ is less than 25" with its mathematical representation.

    (A)  $10(2 + x) < 25$    (B)  $10 + 2x < 25$

    (C)  $10 + 2x > 25$    (D)  $2 + 10x < 25$

    (E)  $10 - 2x < 25$

11. *Multi-Step Problem*    You are saving money to go to summer camp. You need at least $500 to cover all your costs. You can save $20 every week.

    a. Write an inequality to model the situation using $w$ for the number of weeks.

    b. What do the 500 and the 20 represent?

    c. How long will it take you to save enough money?

    d. If you only save $15 a week, how long will it take you to save enough money?

**Algebra 1, Concepts and Skills**
Standardized Test Practice Workbook

# Standardized Test Practice

**For use with pages 30–35**

**TEST TAKING STRATEGY**   **Skip questions that are too difficult for you, and spend no more than a few minutes on each question.**

1. *Multiple Choice*   Choose the correct algebraic translation of "five less than a number."

   Ⓐ  $5 - x$          Ⓑ  $5 \div x$

   Ⓒ  $x + 5$          Ⓓ  $x - 5$

   Ⓔ  $x \div 5$

2. *Multiple Choice*   Choose the correct algebraic translation of "five plus the product of eight and a number."

   Ⓐ  $8(5 + x)$   Ⓑ  $5 + 8x$   Ⓒ  $5 + \dfrac{8}{x}$

   Ⓓ  $8 + 5x$   Ⓔ  $5 + (8 + x)$

3. *Multiple Choice*   Choose the correct algebraic translation of "the quotient of 6 and a number."

   Ⓐ  $6n$          Ⓑ  $\dfrac{6}{n}$

   Ⓒ  $6 + n$          Ⓓ  $3^2 + \dfrac{n}{6}$

   Ⓔ  $6 - n$

4. *Multiple Choice*   Choose the correct algebraic translation of "the product of 12 and a number is 7."

   Ⓐ  $12x = 7$          Ⓑ  $\dfrac{x}{12} = 7$

   Ⓒ  $12 + x = 7$          Ⓓ  $12 = 7x$

   Ⓔ  $\dfrac{12}{x} = 7$

5. *Multiple Choice*   Choose the correct algebraic translation of "thirty-six is the product of $w$ and 12."

   Ⓐ  $36 = w + 12$          Ⓑ  $36 = \dfrac{w}{12}$

   Ⓒ  $w - 12 = 36$          Ⓓ  $36w = 12$

   Ⓔ  $36 = 12w$

6. *Multiple Choice*   Choose the correct algebraic translation of "the difference of seven and a number is less than fourteen."

   Ⓐ  $7 + x < 14$          Ⓑ  $7 + x > 14$

   Ⓒ  $7 - x < 14$          Ⓓ  $7 - x > 14$

   Ⓔ  $\dfrac{7}{x} < 14$

7. *Multiple Choice*   You want to buy cans of juice, which cost $.40 each. You have $5 to spend on juice. Choose the correct model to find the number of cans, $x$, that you can buy.

   Ⓐ  $\dfrac{x}{0.4} \le 5.00$          Ⓑ  $\dfrac{0.4}{x} \le 5.00$

   Ⓒ  $\dfrac{x}{0.4} \ge 5.00$          Ⓓ  $0.4x \ge 5.00$

   Ⓔ  $0.4x \le 5.00$

*Quantitative Comparison*   In Exercises 8 and 9, choose the statement below that is true about the unknown numbers.

   Ⓐ   The solution in column A is greater.

   Ⓑ   The solution in column B is greater.

   Ⓒ   The two solutions are equal.

   Ⓓ   The relationship cannot be determined from the given information.

|     | *Column A* | *Column B* |
| --- | --- | --- |
| 8. | A number decreased by three is six. | Ten times a number is 90. |
| 9. | The product of a number and eight is 32. | The quotient of a number and two is five. |

NAME _____ DATE _____

# *Standardized Test Practice*

**For use with pages 36–41**

**TEST TAKING STRATEGY** **Some questions involve more than one step. Reading too quickly might lead to mistaking the answer to a preliminary step for your final answer.**

*Multiple Choice*  For Exercises 1–5, use the following information and diagram. You are planting a garden with 3 varieties of flowers. Your garden space is 36 inches deep by 72 inches wide. You want to leave 3 inches of space all around the outside, and 3 inches between flower varieties. Assume each flower bed will be the same width.

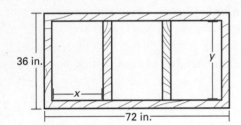

36 in.

*y*

*x*

72 in.

1. Which equation can be used to find the width of each variety of flower bed?

   **A** $3x + 12 = 72$  **B** $3x + 12 = 36$

   **C** $3x + 9 = 72$  **D** $3x + 15 = 72$

   **E** $3x + 9 = 36$

2. How wide can each variety of flower bed be?
   **A** 8 in.  **B** 20 in.
   **C** 21 in.  **D** 19 in.
   **E** 9 in.

3. Which equation can be used to find the depth of each flower bed?

   **A** $6 + 3y = 36$  **B** $3 + y = 36$
   **C** $6 + y = 36$  **D** $6 + y = 72$
   **E** $6 + 3y = 72$

4. How deep can each type of flower planting be?
   **A** 10 in.  **B** 3 in.
   **C** 22 in.  **D** 66 in.
   **E** 30 in.

5. *Multiple Choice*  You have $210 and save $5 a week. Your friend has $270, saves nothing, and spends $10 a week. You want to determine how many weeks it will take before you have as much money as your friend. Which model can be used to represent the situation?

   **A** $210 + 10x = 270 - 5x$

   **B** $210 - 5x = 270 + 10x$

   **C** $210 + 5x + 270 - 10x = 0$

   **D** $210 + 5x = 270 - 10x$

   **E** $210 - 10x = 270 + 5x$

6. *Multi-Step Problem*  Two cross-country skiers leave the ski lodge at the same time, and travel the same distance. The first skier returns to the lodge after $2\frac{1}{2}$ hours and averaged 8 miles per hour. The second skier returns 195 minutes later.

   **a.** Draw a diagram modeling the situation.

   **b.** Find the distance traveled by the skiers.

   **c.** Find the second skier's speed.

   **d.** The second skier had to stop for 15 minutes to wax his skis. What was the speed of this skier if you consider only skiing time?

NAME _____    DATE _____

# *Standardized Test Practice*

For use with pages 42–47

**TEST TAKING STRATEGY    Work as fast as you can through the easier problems, but not so fast that you are careless.**

*Multiple Choice*   In Exercises 1–4, use the following information and graph to determine the correct answer.

Australia has the highest incidence of pet ownership in the world, with 66% of households owning a pet.

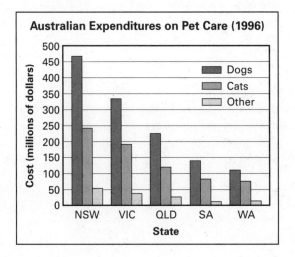

1. Which state has the highest total spending on pet care?

   Ⓐ NSW        Ⓑ VIC        Ⓒ QLD

   Ⓓ SA         Ⓔ WA

2. Which two states spent about the same amount of money on "other" pet care?

   Ⓐ NSW and VIC      Ⓑ VIC and QLD

   Ⓒ QLD and SA       Ⓓ SA and WA

   Ⓔ WA and QLD

3. What is the approximate total expenditure on dog and cat care in SA?

   Ⓐ $345 million     Ⓑ $710 million

   Ⓒ $220 million     Ⓓ $525 million

   Ⓔ $150 million

4. Which state spent approximately $150 million on cat and other pet expenditures?

   Ⓐ NSW        Ⓑ VIC        Ⓒ QLD

   Ⓓ SA         Ⓔ WA

5. *Multiple Choice*   In which three years did home ownership increase?

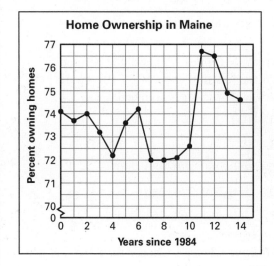

   Ⓐ 1984, 1985, 1986   Ⓑ 1987, 1988, 1989

   Ⓒ 1996, 1997, 1998   Ⓓ 1986, 1987, 1988

   Ⓔ 1993, 1994, 1995

*Quantitative Comparison*   In Exercises 6–8, refer to the graph in Exercise 5 and choose the statement that is true about the given numbers.

   Ⓐ   The number in column A is greater.

   Ⓑ   The number in column B is greater.

   Ⓒ   The two numbers are equal.

   Ⓓ   The relationship cannot be determined from the given information.

| | *Column A* | *Column B* |
|---|---|---|
| **6.** | ownership rate in 1987 | ownership rate in 1990 |
| **7.** | ownership rate in 1990 | ownership rate in 1994 |
| **8.** | ownership rate in 1991 | ownership rate in 1992 |

# *Standardized Test Practice*

For use with pages 48–54

**TEST TAKING STRATEGY** **If you can, check your answer using a method that is different from the one you used originally, to avoid making the same mistake twice.**

1. *Multiple Choice* Choose the correct domain for the function $t = 35 + 5w$ where $0 \le w \le 5$ and $w$ is an integer.

   Ⓐ 0, 1, 2, 3, 4, 5

   Ⓑ 1, 2, 3, 4, 5

   Ⓒ 35, 40, 45, 50, 55

   Ⓓ 40, 45, 50, 55

   Ⓔ 40, 41, 42, 43, 44, 45

2. *Multiple Choice* Choose the correct range for the function $y = 250 + 10x$ where $0 \le x \le 4$ and $x$ is an integer.

   Ⓐ 0, 1, 2, 3, 4

   Ⓑ 0, 1, 2, 3

   Ⓒ 250, 260, 270, 280, 290

   Ⓓ 250, 260, 270, 280

   Ⓔ 260, 261, 262, 263, 264

3. *Multiple Choice* Choose the table that does not represent a function.

   Ⓐ
   | Input | 0 | 1 | 2 | 3 | 4 |
   |---|---|---|---|---|---|
   | Output | 10 | 15 | 20 | 25 | 30 |

   Ⓑ
   | Input | 1 | 2 | 3 | 4 | 5 |
   |---|---|---|---|---|---|
   | Output | 3 | 3 | 5 | 5 | 7 |

   Ⓒ
   | Input | 0 | 0 | 1 | 1 | 2 |
   |---|---|---|---|---|---|
   | Output | 7 | 9 | 11 | 13 | 15 |

   Ⓓ
   | Input | 0 | 1 | 2 | 3 | 4 |
   |---|---|---|---|---|---|
   | Output | 5 | 10 | 15 | 30 | 90 |

   Ⓔ
   | Input | 1 | 2 | 3 | 4 | 5 |
   |---|---|---|---|---|---|
   | Output | 6 | 12 | 18 | 24 | 30 |

4. *Multiple Choice* Choose the equation that describes the function containing all of the points shown in the table.

   | Input | 0 | 1 | 2 | 3 | 4 | 5 |
   |---|---|---|---|---|---|---|
   | Output | 7 | 15 | 23 | 31 | 39 | 47 |

   Ⓐ $y = 12 + x^2$   Ⓑ $y = 7 + 2x^2$

   Ⓒ $y = 4x + 7$   Ⓓ $y = 16 - 2x$

   Ⓔ $y = 8x + 7$

5. *Multiple Choice* A bag of puppy food directs you to feed $\frac{1}{2}$ cup for every 10 pounds the puppy weighs. Choose the correct representation of cups, $c$, as a function of body weight, $w$.

   Ⓐ $c = \frac{1}{2}(10w)$   Ⓑ $c = \frac{1}{2}\left(\frac{w}{10}\right)$

   Ⓒ $c = \frac{1}{2} + w$   Ⓓ $w = \frac{1}{2} + 10c$

   Ⓔ $c = \frac{1}{2}(w - 10)$

6. *Multi-Step Problem* Your high school band is selling sub sandwiches for a fund raiser. It costs $1.25 to make each sandwich. They sell them for $2.50 each.

   a. Write a function that determines the profit, $p$, from the sale based on $x$, the number of sandwiches sold.

   b. Make a table of input $x$ and output $p$ for $x = 50, 100, 150,$ and 200.

   c. The band must pay a $75 fee to use the cafeteria to assemble the sandwiches. Write a function that includes this cost.

**Algebra 1, Concepts and Skills**
Standardized Test Practice Workbook

Chapter 1

NAME _____ DATE _____

# *Standardized Test Practice*

For use with pages 65–70

**TEST TAKING STRATEGY**   **If you can, check an answer using a method that is different from the one you used originally, to avoid making the same mistake twice.**

1. *Multiple Choice*   Choose the number that best represents the graph.

   (**A**)  $\frac{1}{2}$      (**B**)  $-\frac{1}{2}$      (**C**)  $\frac{1}{4}$

   (**D**)  $-\frac{1}{4}$      (**E**)  $-1$

2. *Multiple Choice*   Choose the correct ordering (in increasing order) of the following numbers: $2, -0.5, 0, -1, 0.75$.

   (**A**)  $0, -0.5, 0.75, -1, 2$

   (**B**)  $-0.5, -1, 0, 0.75, 2$

   (**C**)  $-0.5, -1, 0.75, 0, 2$

   (**D**)  $-1, -0.5, 0, 0.75, 2$

   (**E**)  $-1, -0.5, 0.75, 0, 2$

3. *Multiple Choice*   Choose the correct ordering (in increasing order) of the following numbers: $0, 6, -3, \frac{1}{2}, -0.5, -\frac{3}{4}$.

   (**A**)  $0, -3, -0.5, \frac{1}{2}, -\frac{3}{4}, 6$

   (**B**)  $-3, -0.5, -\frac{3}{4}, 0, \frac{1}{2}, 6$

   (**C**)  $0, -0.5, \frac{1}{2}, -\frac{3}{4}, -3, 6$

   (**D**)  $-3, -\frac{3}{4}, -0.5, \frac{1}{2}, 0, 6$

   (**E**)  $-3, -\frac{3}{4}, -0.5, 0, \frac{1}{2}, 6$

*Multiple Choice*   For Exercises 4 and 5, refer to the table showing high temperatures in January for one week in Pittsburgh, PA.

| Date | 1/3 | 1/4 | 1/5 | 1/6 | 1/7 | 1/8 | 1/9 |
|------|-----|-----|-----|-----|-----|-----|-----|
| °C | 10° | 5° | −5° | −10° | 3° | 0° | −2° |

4. What was the coldest temperature recorded?

   (**A**)  $0°$      (**B**)  $3°$      (**C**)  $-2°$

   (**D**)  $-5°$      (**E**)  $-10°$

5. Which days had high temperatures above freezing (0°C)?

   (**A**)  Jan. 3, 4, 7      (**B**)  Jan. 3, 4, 7, 8

   (**C**)  Jan. 3, 4, 5, 6, 7, 9      (**D**)  Jan. 3, 4

   (**E**)  Jan. 5, 6, 9

6. *Multiple Choice*   Choose the inequality that correctly compares $-4$ and $-6$.

   (**A**)  $-4 = -6$      (**B**)  $-4 < -6$

   (**C**)  $-6 < -4$      (**D**)  $-6 > -4$

   (**E**)  Both B and C

7. *Multiple Choice*   Choose the real number that is greater than 0.01.

   (**A**)  –0.1      (**B**)  –0.001      (**C**)  0.1

   (**D**)  0      (**E**)  0.001

*Quantitative Comparison*   In Exercises 8–10, choose the statement that is true about the given numbers.

   (**A**)  The number in column A is greater.

   (**B**)  The number in column B is greater.

   (**C**)  The two numbers are equal.

   (**D**)  The relationship cannot be determined from the information given.

| | Column A | Column B |
|---|---|---|
| 8. | 4.9 | −4.2 |
| 9. | −7.2 | −2.7 |
| 10. | 2.11 | 2.1 |

11. *Multiple Choice*   Choose the inequality that correctly compares $-\frac{1}{11}$ and $-\frac{1}{12}$.

   (**A**)  $-\frac{1}{11} < -\frac{1}{12}$      (**B**)  $-\frac{1}{11} > -\frac{1}{12}$

   (**C**)  $-\frac{1}{11} = -\frac{1}{12}$      (**D**)  $-\frac{1}{12} < -\frac{1}{11}$

   (**E**)  $-\frac{1}{12} \le -\frac{1}{11}$

NAME _____ DATE _____

# Standardized Test Practice

**For use with pages 71–76**

**TEST TAKING STRATEGY**  **Spend no more than a few minutes on each question.**

**1. *Multiple Choice***   What is the opposite of 5?

 (A)  $|5|$    (B)  $-5$    (C)  $|-5|$

 (D)  $-|-5|$    (E)  Both B and D

**2. *Multiple Choice***   Solve the equation: $|x| = 10$.

 (A)  $-10$     (B)  $10$

 (C)  $-10$ and $10$    (D)  $|-10|$ and $|10|$

 (E)  no solution

**3. *Multiple Choice***   Solve the equation: $|x| = -10$.

 (A)  $-10$     (B)  $10$

 (C)  $-10$ and $10$    (D)  $|-10|$ and $|10|$

 (E)  no solution

**4. *Multiple Choice***   Complete the statement: The absolute value of a number is always _____?_____.

 (A)  increasing    (B)  decreasing

 (C)  positive    (D)  negative

 (E)  None of these

**5. *Multiple Choice***   Complete the statement: The opposite of a number is always _____?_____.

 (A)  larger than the number    (B)  smaller than the number

 (C)  positive    (D)  negative

 (E)  None of these

**6. *Multiple Choice***   Evaluate the expression $|2| - |-2|$.

 (A)  $-2$     (B)  $0$

 (C)  $2$     (D)  $4$

 (E)  $-4$

**7. *Multiple Choice***   Evaluate the expression $-|-2|$.

 (A)  $-2$    (B)  $2$    (C)  $0$

 (D)  $1$    (E)  $-1$

**8. *Multiple Choice***   Evaluate the expression $|-7| + 3$.

 (A)  $-4$    (B)  $4$    (C)  $3$

 (D)  $10$    (E)  $-10$

**9. *Multiple Choice***   The velocity of an airplane during takeoff is _____?_____.

 (A)  zero    (B)  negative

 (C)  positive    (D)  constant

 (E)  none of these

***Quantitative Comparison***   In Exercises 10–12, choose the statement that is true about the given numbers.

 (A)  The number in column A is greater.

 (B)  The number in column B is greater.

 (C)  The two numbers are equal.

 (D)  The relationship cannot be determined from the information given.

| | Column A | Column B |
|---|---|---|
| **10.** | $-|3 + 2|$ | $|3 + 2|$ |
| **11.** | $|4| + 2$ | $|-4| + 2$ |
| **12.** | $|22 - 6|$ | $|22| - 6$ |

**Algebra 1, Concepts and Skills**
Standardized Test Practice Workbook

NAME _____ DATE _____

# Standardized Test Practice

**For use with pages 78–83**

**TEST TAKING STRATEGY**  **Spend no more than a few minutes on each question.**

1. **Multiple Choice**  Find the sum of $3 + (-4)$.
   - Ⓐ 7
   - Ⓑ $-7$
   - Ⓒ $-1$
   - Ⓓ 1
   - Ⓔ 12

2. **Multiple Choice**  Find the sum of $-6 + (-18)$.
   - Ⓐ $-12$
   - Ⓑ 12
   - Ⓒ 24
   - Ⓓ $-24$
   - Ⓔ 3

3. **Multiple Choice**  Find the sum of $-5 + 8 + (-2)$.
   - Ⓐ 15
   - Ⓑ $-15$
   - Ⓒ 1
   - Ⓓ $-1$
   - Ⓔ 11

4. **Multiple Choice**  Find the sum of $2.3 + (-0.5) + (-1.2) + (-0.7)$.
   - Ⓐ $-0.1$
   - Ⓑ 0.1
   - Ⓒ 0.98
   - Ⓓ 4.7
   - Ⓔ $-4.7$

5. **Multiple Choice**  Evaluate the expression $5 + x + (-3)$ for $x = -2$.
   - Ⓐ 6
   - Ⓑ $-6$
   - Ⓒ 0
   - Ⓓ 1
   - Ⓔ $-1$

**Multiple Choice**  In Exercises 6–8, use the table, which shows the average daily high temperature and the actual temperature of a city in degrees Celsius.

| Date | Avg. High | Act. High |
|---|---|---|
| Jan. 8 | 2 | 4 |
| Jan. 9 | 0 | 5 |
| Jan. 10 | $-1$ | $-3$ |
| Jan. 11 | $-3$ | $-11$ |
| Jan. 12 | 1 | $-8$ |
| Jan. 13 | 3 | $-2$ |
| Jan. 14 | 5 | 0 |

6. On which date was the actual high temperature furthest from the average temperature?
   - Ⓐ Jan. 11
   - Ⓑ Jan. 13
   - Ⓒ Jan. 8
   - Ⓓ Jan. 12
   - Ⓔ Jan. 9

7. On which dates were the actual temperatures 5 degrees different from the average high temperatures?
   - Ⓐ Jan. 9, 11
   - Ⓑ Jan. 9, 13
   - Ⓒ Jan. 9, 14
   - Ⓓ Jan. 13, 14
   - Ⓔ Jan. 9, 13, 14

8. When was the actual temperature closest to the average temperature?
   - Ⓐ Jan. 8
   - Ⓑ Jan. 10
   - Ⓒ Jan. 13
   - Ⓓ Jan. 8, 10
   - Ⓔ Jan. 8, 10, 13

*Quantitative Comparison*  In Exercises 9–12, choose the statement that is the true about the given quantities.
   - Ⓐ The quantity in column A is greater.
   - Ⓑ The quantity in column B is greater.
   - Ⓒ The two quantities are equal.
   - Ⓓ The relationship cannot be determined from the information given.

|  | Column A | Column B |
|---|---|---|
| 9. | $-3 + (-2)$ | $\lvert 3 + 2 \rvert$ |
| 10. | $-\frac{1}{2} + 1\frac{2}{3}$ | $-\frac{1}{6} + 2$ |
| 11. | $8.2 + (-4.6) + 0.4$ | $-5.1 + 7.9 + (-2)$ |
| 12. | $(-2) + (-3) + (-7)$ | $8 + (-11) + (-9)$ |

**Algebra 1, Concepts and Skills**
Standardized Test Practice Workbook

NAME _____ DATE _____

# *Standardized Test Practice*

For use with pages 86–91

**TEST TAKING STRATEGY**   **Work as fast as you can through the easier problems, but not so fast that you are careless.**

1. *Multiple Choice*   Find the difference of $12 - 18$.
   (A)  6       (B)  $-6$       (C)  30
   (D)  $-30$       (E)  $\frac{12}{18}$

2. *Multiple Choice*   Evaluate the expression $-7 - 10 + 2$.
   (A)  5       (B)  15       (C)  $-15$
   (D)  $-19$       (E)  1

3. *Multiple Choice*   Evaluate the expression $-6 - (-12) + 8$.
   (A)  26       (B)  2       (C)  $-2$
   (D)  14       (E)  $-10$

4. *Multiple Choice*   Evaluate: $-\frac{1}{2} - \frac{2}{3} - \left(-\frac{5}{8}\right)$.
   (A)  $\frac{13}{24}$       (B)  $-\frac{13}{24}$       (C)  $\frac{2}{3}$
   (D)  $-\frac{19}{24}$       (E)  $-1\frac{19}{24}$

5. *Multiple Choice*   Choose the correct terms of the expression $-5x - 6$.
   (A)  $-5$       (B)  $-6$
   (C)  $-5, -6$   (D)  $-5x, -6$
   (E)  $-5x, 6$

6. *Multiple Choice*   Choose the correct terms of the expression $9 - 15x + 12y$.
   (A)  $9, -15, 12$   (B)  $9, 15, 12$
   (C)  $9, 15x, 12y$   (D)  $9, -15x, 12y$
   (E)  $9, x, y$

7. *Multiple Choice*   Choose the outputs of the function $y = -x + 5$ for these values of $x$: $-2, -1, 0$, and 1.
   (A)  7, 6, 5, 4   (B)  3, 4, 5, 6
   (C)  7, 6, 5, 6   (D)  3, 4, 5, 4
   (E)  $-3, -4, -5, -6$

*Multiple Choice*   For Exercises 8 and 9, use the table, which shows the average price of regular unleaded gasoline in Cleveland, Ohio for five weeks.

| Week | 1 | 2 | 3 | 4 | 5 |
|---|---|---|---|---|---|
| dollars/gallon | $1.17 | $1.29 | $1.09 | $1.13 | $1.07 |

8. What is the change in the price between week 1 and week 3?
   (A)  $-\$.20$       (B)  $\$.20$       (C)  $\$.08$
   (D)  $\$.09$       (E)  $-\$.08$

9. What is the greatest positive one week change?
   (A)  $\$.20$       (B)  $\$.12$       (C)  $\$.04$
   (D)  $\$.06$       (E)  $\$.10$

10. *Multiple Choice*   There was 26.2 inches of snow on the ground on Monday. On Tuesday 3 more inches fell. On Wednesday 4.2 inches melted off. On Thursday 2.6 inches melted. Friday 1.2 inches accumulated. How deep was the snow by Friday night?
    (A)  26.8 in       (B)  19.2 in.
    (C)  23.6 in.       (D)  21.2 in.
    (E)  15.2 in.

11. *Multi-Step Problem*   Your science class is doing a fall bird count. Each day for a week the birds sighted by class members are counted. The results are as follows.

| Mon | Tues. | Wed. | Thurs. | Fri. |
|---|---|---|---|---|
| 52 | 50 | 61 | 59 | 48 |

   a. Find the change in the number of birds counted from each day to the next.

   b. What does a negative value represent?

   c. What does a positive value represent?

**Algebra 1, Concepts and Skills**
Standardized Test Practice Workbook

NAME _____ DATE _____

# *Standardized Test Practice*

For use with pages 93–98

**TEST TAKING STRATEGY** Spend no more than a few minutes on each question.

1. *Multiple Choice* Find the product $(-8)(2)(-2)$.
   - Ⓐ 32
   - Ⓑ −32
   - Ⓒ −8
   - Ⓓ 8
   - Ⓔ 18

2. *Multiple Choice* Find the product $\left(-\frac{1}{2}\right)(-6)\left(-\frac{1}{3}\right)$.
   - Ⓐ $6\frac{1}{6}$
   - Ⓑ $-6\frac{1}{6}$
   - Ⓒ 1
   - Ⓓ −1
   - Ⓔ $-\frac{6}{5}$

3. *Multiple Choice* Simplify $(-3)(-x)(4)$.
   - Ⓐ $-12x$
   - Ⓑ $12x$
   - Ⓒ $x$
   - Ⓓ $-1x$
   - Ⓔ $-12 - x$

4. *Multiple Choice* Simplify $-(-x)(6)\left(\frac{1}{2}x\right)$.
   - Ⓐ $3x^2$
   - Ⓑ $-3x^2$
   - Ⓒ $3 + 2x$
   - Ⓓ $-3 + 2x$
   - Ⓔ $\frac{7}{2}x^2$

5. *Multiple Choice* The equation $a \cdot b = b \cdot a$ represents which property of multiplication?
   - Ⓐ commutative property
   - Ⓑ associative property
   - Ⓒ identity property
   - Ⓓ property of zero
   - Ⓔ property of opposites

6. *Multiple Choice* Evaluate the expression $(2x)(-3)(-1)$ for $x = -4$.
   - Ⓐ −24
   - Ⓑ 24
   - Ⓒ 18
   - Ⓓ −18
   - Ⓔ −6

7. *Multiple Choice* Evaluate the expression $(3x)^2 - 8x$ when $x = -3$.
   - Ⓐ 12
   - Ⓑ −57
   - Ⓒ 105
   - Ⓓ 60
   - Ⓔ 57

8. *Multiple Choice* A hawk dives towards the ground to catch a rabbit. It descends with a velocity of $-10.2$ feet per second. What would be the change in position in 5 seconds?
   - Ⓐ −15.2 ft
   - Ⓑ 2.04 ft
   - Ⓒ −2.04 ft
   - Ⓓ 51 ft
   - Ⓔ −51 ft

9. *Multiple Choice* A store is marking down its winter clothing for an end of season sale. A fleece jacket that cost the store $50 is now being sold for $43. The store has 30 jackets left to sell. If all the jackets are sold during the sale, how much will the store lose?
   - Ⓐ $7
   - Ⓑ $37
   - Ⓒ $210
   - Ⓓ $1290
   - Ⓔ $1500

10. *Multi-Step Problem* Your sister lends you $75 to buy a video game. You agree to work off your loan by doing her chores for $3 an hour.
    a. Write an equation that you can use to determine the amount of money, $m$, you still owe your sister after working $h$ hours.
    b. How much do you owe after 12 hours of chores?
    c. How many hours of chores must you complete to pay for the $75 loan?

**TEST TAKING STRATEGY   Skip questions that are too difficult for you.**

1. *Multiple Choice*   Use the distributive property to rewrite $6(x - 3)$ without parentheses.

   **A**  $6x + 3$        **B**  $6x - 18$

   **C**  $6x + 18$       **D**  $6x - 3$

   **E**  $x - 18$

2. *Multiple Choice*   Use the distributive property to rewrite $(3y - 9)(8)$ without parentheses.

   **A**  $24y - 72$      **B**  $24y + 72$
   **C**  $11y$           **D**  $3y - 72$
   **E**  $24y - 9$

3. *Multiple Choice*   Choose the expression that is equivalent to $8x - 4y$.

   **A**  $(8x - 4)y$     **B**  $8(x - 2y)$

   **C**  $4(2x - y)$     **D**  $(2x + y)(4)$

   **E**  $2(4x - y)$

4. *Multiple Choice*   Choose the expression that is equivalent to $\dfrac{x}{3} - 1$.

   **A**  $\frac{1}{6}(2x - 6)$        **B**  $(x - 6)\frac{1}{3}$

   **C**  $(x - 1)\left(\frac{1}{3}\right)$     **D**  $-(x - 6)\frac{1}{3}$

   **E**  $-\frac{1}{6}(2x - 6)$

5. *Multiple Choice*   Use the distributive property to rewrite $-4(5x - 2.5)$ without parentheses.

   **A**  $-20x + 2.5$    **B**  $-9x + 6.5$

   **C**  $-9x - 6.5$     **D**  $-20x - 10$

   **E**  $-20x + 10$

6. *Multiple Choice*   Use mental math to simplify the expression: $6(9.9)$.

   **A**  $59.6$          **B**  $59.4$

   **C**  $58.8$          **D**  $56.9$

   **E**  $54.9$

7. *Multiple Choice*   You are buying 6 pairs of socks. The price of each pair of socks is $5.95. What expression could you use to mentally calculate the total cost of the socks?

   **A**  $6(6) + 0.05$     **B**  $6(6) - 0.05$

   **C**  $6(6 - 0.05)$     **D**  $6(6 + 0.05)$

   **E**  $6 - 6(0.05)$

8. *Multi-Step Problem*   Use the area model shown below.

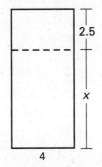

   **a.** Find two expressions for the area of the rectangle.

   **b.** Are the two expressions equal?

# LESSON
## 2.7

NAME _____ DATE _____

# *Standardized Test Practice*

For use with pages 107–112

**TEST TAKING STRATEGY**   **Spend no more than a few minutes on each question.**

**1.** *Multiple Choice*   In the expression $3s^5 - 2s^2 + 1$, what is the coefficient of the $s^2$ term?

  Ⓐ  3       Ⓑ  $-3$

  Ⓒ  $-2$      Ⓓ  2

  Ⓔ  1

**2.** *Multiple Choice*   Which expression is simplified?

  Ⓐ  $2 + x^2 - 1$    Ⓑ  $y^3 + 2y - 5y^3$

  Ⓒ  $4z - 4z + z^5$    Ⓓ  $1 - v^2 - v$

  Ⓔ  $w + 5w^3 + 2w$

**3.** *Multiple Choice*   Simplify the expression $x - x^2 + 3x^2 + 5x$.

  Ⓐ  $2x^2 + 6x$     Ⓑ  $6x - x^2 + 3x^2$

  Ⓒ  $x + 2x^2 + 5x$ Ⓓ  $2x^2 + 5x$

  Ⓔ  $6x^2 + 2x$

**4.** *Multiple Choice*   Simplify the expression $10x - 3 + 4 - 2x$.

  Ⓐ  $8x - 1$     Ⓑ  $12x + 1$

  Ⓒ  $12x + 7$    Ⓓ  $8x + 1$

  Ⓔ  $8x - 7$

**5.** *Multiple Choice*   Simplify the expression $2x^2 - 5 + 3x^2 - 4x + 6$.

  Ⓐ  $x + 1$      Ⓑ  $6x^2 - 4x + 1$

  Ⓒ  $5x^2 - 4x + 11$ Ⓓ  $5x^2 - 4x + 1$

  Ⓔ  $2x^2 + 1$

**6.** *Multiple Choice*   Apply the distributive property, then simplify: $(4x - 2)(-5) + 3x$.

  Ⓐ  $-17x - 10$   Ⓑ  $-23x - 10$

  Ⓒ  $-17x + 10$   Ⓓ  $-23x + 10$

  Ⓔ  $2x - 7$

**7.** *Multiple Choice*   Apply the distributive property, then simplify: $6y - 2(3y - 8) + 2y$.

  Ⓐ  $2y - 16$    Ⓑ  $16 + 2y$

  Ⓒ  $14y - 16$   Ⓓ  $14y + 16y$

  Ⓔ  $2y - 8$

**8.** *Multiple Choice*   Write and simplify an expression modeling the perimeter of the rectangle.

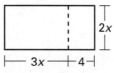

  Ⓐ  $5x + 4$      Ⓑ  $10x + 8$

  Ⓒ  $6x + 12$    Ⓓ  $6x^2 + 8x$

  Ⓔ  $8x + 8$

**9.** *Multi-Step Problem*   You and your friends are going on a 4 hour canoe trip. The creek is low so you will have to portage, or carry, your canoe for some of the trip. You can travel 6 miles per hour in your canoe, and 0.6 mile per hour when portaging.

  **a.** Let $t$ represent the time in hours you portage your canoe. Which function can you use to find the total distance traveled on your canoe trip?

    Ⓐ  $T = 0.6t + 6(4 - t)$

    Ⓑ  $T = \dfrac{0.6}{t} + \dfrac{6}{4 - t}$

    Ⓒ  $T = \dfrac{t}{0.6} + \dfrac{4 - t}{6}$

  **b.** If you portage your canoe for 30 min, or 0.5 hours, how far did you travel?

Chapter 2

NAME _____ DATE _____

# *Standardized Test Practice*

For use with pages 113–118

**TEST TAKING STRATEGY**   **Work as fast as you can through the easier problems, but not so fast that you make careless mistakes.**

**1.** *Multiple Choice*   The expression $b \div a$ is equivalent to ___?___.

  **(A)** $b \cdot \dfrac{a}{1}$     **(B)** $b \cdot \dfrac{1}{a}$     **(C)** $a \cdot \dfrac{b}{1}$

  **(D)** $a \cdot \dfrac{1}{b}$     **(E)** $\dfrac{a}{b}$

**2.** *Multiple Choice*   Find the quotient.
$-39 \div 3\frac{1}{4} = ?$

  **(A)** $-126\frac{3}{4}$   **(B)** $126\frac{3}{4}$   **(C)** $-12$

  **(D)** $12$     **(E)** $-\frac{4}{3}$

**3.** *Multiple Choice*   Find the quotient.
$-\frac{2}{3} \div \left(-\frac{3}{4}\right) = ?$

  **(A)** $\frac{8}{9}$     **(B)** $-\frac{8}{9}$     **(C)** $\frac{1}{2}$

  **(D)** $-\frac{1}{2}$     **(E)** $\frac{6}{12}$

**4.** *Multiple Choice*   Simplify the expression
$\dfrac{24x - 8}{4}$.

  **(A)** $24x - 2$     **(B)** $6x - 2$

  **(C)** $6x - 8$     **(D)** $6x + 2$

  **(E)** $96x - 32$

**5.** *Multiple Choice*   Simplify the expression
$\dfrac{-56}{12} \div \dfrac{8}{6}$.

  **(A)** $\dfrac{-56}{9}$   **(B)** $\dfrac{-2}{7}$   **(C)** $\dfrac{2}{7}$

  **(D)** $\dfrac{-7}{2}$     **(E)** $\dfrac{7}{2}$

**6.** *Multiple Choice*   Simplify the expression
$\dfrac{14x - 21}{7}$.

  **(A)** $2x - 3$   **(B)** $7x - 14$

  **(C)** $2x + 3$   **(D)** $2x - 21$

  **(E)** $14x - 3$

**7.** *Multiple Choice*   Simplify the expression
$-64 \div \dfrac{16}{5}$.

  **(A)** $-20$   **(B)** $20$   **(C)** $-9$

  **(D)** $-204.8$   **(E)** $9$

**8.** *Multiple Choice*   Choose the value of $x$ which cannot be in the domain of
$y = \dfrac{8}{5 - x}$.

  **(A)** $0$   **(B)** $5$   **(C)** $-5$

  **(D)** $3$   **(E)** $-3$

**9.** *Multiple Choice*   A roller coaster descends from its highest point of 200 feet in 5 seconds. What is the velocity?

  **(A)** $-40$ ft/sec   **(B)** $40$ ft/sec

  **(C)** $0.025$ ft/sec   **(D)** $-0.025$ ft/sec

  **(E)** $-1000$ ft/sec

*Quantitative Comparison*   In Exercises 10–13, choose the statement that is true about the given quantities when $a = -2$ and $b = 3$.

  **(A)** The quantity in column A is greater.

  **(B)** The quantity in column B is greater.

  **(C)** The two quantities are equal.

  **(D)** The relationship cannot be determined from the information given.

| | Column A | Column B |
|---|---|---|
| **10.** | $\dfrac{a - 3}{4}$ | $\dfrac{18 - b}{b}$ |
| **11.** | $\dfrac{-ab - 8}{a}$ | $\dfrac{a}{b} \div \dfrac{b}{a}$ |
| **12.** | $\dfrac{b - a}{3}$ | $\dfrac{ab}{b - a}$ |
| **13.** | $\dfrac{3a - 5b}{5}$ | $\dfrac{5a - b}{a}$ |

**Algebra 1, Concepts and Skills**
Standardized Test Practice Workbook

NAME _____ DATE _____

# *Standardized Test Practice*

**For use with pages 132–137**

**TEST TAKING STRATEGY**  **Go back and check as much of your work as you can.**

1. *Multiple Choice*  Choose the inverse operation of "Subtract $-17$."
   - Ⓐ  Multiply by 17.
   - Ⓑ  Add $-17$.
   - Ⓒ  Multiply by $-17$.
   - Ⓓ  Add 17.
   - Ⓔ  Divide by $-17$.

2. *Multiple Choice*  Choose the inverse operation of "Add 8."
   - Ⓐ  Multiply by 8.
   - Ⓑ  Subtract 8.
   - Ⓒ  Divide by 8.
   - Ⓓ  Subtract $-8$.
   - Ⓔ  Multiply by $-8$.

3. *Multiple Choice*  Solve $x - 6 = 10$.
   - Ⓐ  4
   - Ⓑ  $-4$
   - Ⓒ  16
   - Ⓓ  $-16$
   - Ⓔ  60

4. *Multiple Choice*  Solve $14 = y + 2$.
   - Ⓐ  16
   - Ⓑ  $-16$
   - Ⓒ  $-12$
   - Ⓓ  12
   - Ⓔ  7

5. *Multiple Choice*  Solve $8 + x = -10$.
   - Ⓐ  $-18$
   - Ⓑ  18
   - Ⓒ  $-2$
   - Ⓓ  2
   - Ⓔ  1.2

6. *Multiple Choice*  Solve $3 = 10 + x$.
   - Ⓐ  13
   - Ⓑ  $-13$
   - Ⓒ  $-7$
   - Ⓓ  $-3$
   - Ⓔ  7

7. *Multiple Choice*  There are 22 students in a Spanish class. This is an increase of 8 students from last year. Which equation could be used to determine the number of students in last year's Spanish class?
   - Ⓐ  $22 + x = 8$
   - Ⓑ  $x - 8 = 22$
   - Ⓒ  $22 + x = -8$
   - Ⓓ  $x + 8 = 22$
   - Ⓔ  $22 + 8 = x$

8. *Multiple Choice*  There was 2.6 inches of rainfall today. The total precipitation for the month is now 8.2 inches. Which equation could be used to determine the amount of precipitation before today?
   - Ⓐ  $x - 2.6 = 8.2$
   - Ⓑ  $2.6 + 8.2 = x$
   - Ⓒ  $x + 2.6 = 8.2$
   - Ⓓ  $8.2 + x = 2.6$
   - Ⓔ  $2.6 - x = 8.2$

9. *Multiple Choice*  Find the length of the side $x$ of the triangle. The perimeter is 14.
   - Ⓐ  3
   - Ⓑ  4
   - Ⓒ  5
   - Ⓓ  6
   - Ⓔ  7

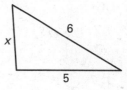

10. *Multiple Choice*  A pet store marks up the price of its puppies by $150. It sells them for $500 each. How much did the store originally pay for each puppy?
   - Ⓐ  $650
   - Ⓑ  $350
   - Ⓒ  $250
   - Ⓓ  $150
   - Ⓔ  $500

*Quantitative Comparison*  In Exercises 11–14, choose the statement below that is true about the given numbers.
   - Ⓐ  The solution in column A is greater.
   - Ⓑ  The solution in column B is greater.
   - Ⓒ  The two solutions are equal.
   - Ⓓ  The relationship cannot be determined from the given information.

| | Column A | Column B |
|---|---|---|
| 11. | $x + 2 = 6$ | $y - 4 = -6$ |
| 12. | $x + |-3| = 3$ | $6 = y + 4$ |
| 13. | $|x| = 8$ | $y - 3 = 4$ |
| 14. | $-6 = x + 3$ | $-2 = y - (-8)$ |

# Standardized Test Practice

For use with pages 138–143

**TEST TAKING STRATEGY**  **Before you give up on a question, try to eliminate some of your choices so you can make an educated guess.**

1. *Multiple Choice*  Choose the inverse operation of "multiply by −6."

   Ⓐ Multiply by 6.  Ⓑ Divide by 6.

   Ⓒ Divide by −6.  Ⓓ Add 6.

   Ⓔ Add −6.

2. *Multiple Choice*  Which one of these steps can by used to solve the equation $\frac{2}{3}x = 8$?

   **I.** Multiply by $\frac{2}{3}$.  **II.** Multiply by $\frac{3}{2}$.

   **III.** Divide by $\frac{2}{3}$.  **IV.** Divide by $\frac{3}{2}$.

   Ⓐ II only  Ⓑ III only

   Ⓒ I and III  Ⓓ II and III

   Ⓔ II and IV

3. *Multiple Choice*  Solve $18 = \dfrac{x}{-3}$.

   Ⓐ −6  Ⓑ 6  Ⓒ −54

   Ⓓ 54  Ⓔ 21

4. *Multiple Choice*  Solve $-\frac{1}{3}y = \frac{4}{5}$.

   Ⓐ $-2\frac{2}{5}$  Ⓑ $2\frac{2}{5}$  Ⓒ $-\frac{5}{12}$

   Ⓓ $\frac{5}{12}$  Ⓔ $1\frac{2}{15}$

5. *Multiple Choice*  If $-0.3x = 12$, then $x = \underline{\ ?\ }$.

   Ⓐ −40  Ⓑ 40  Ⓒ 3.6

   Ⓓ −3.6  Ⓔ 12.3

6. *Multiple Choice*  Each week the average household reads 5 pounds of newspapers. You are recycling 50 pounds of newspapers. Which equation could be use to estimate the number of weeks during which your household read the papers?

   Ⓐ $50x = 5$

   Ⓑ $5x = 50$

   Ⓒ $5 + x = 50$

   Ⓓ $50 + x = 5$

   Ⓔ $50 - x = 5$

7. *Multiple Choice*  You and two of your friends went to a restaurant for lunch and divided the bill evenly. If your cost was $7.50, find the total bill.

   Ⓐ $2.50  Ⓑ $10.50  Ⓒ $30

   Ⓓ $20  Ⓔ $22.50

8. *Multiple Choice*  If $a = b$ and $c \neq 0$, then $\dfrac{a}{c} = \dfrac{b}{c}$ represents which property of equality?

   Ⓐ Addition property of equality

   Ⓑ Subtraction property of equality

   Ⓒ Multiplication property of equality

   Ⓓ Division property of equality

   Ⓔ Reciprocal property of equality

*Quantitative Comparison*  In Exercises 9–11, choose the statement that is the true about the given numbers.

   Ⓐ The solution in column A is greater.

   Ⓑ The solution in column B is greater.

   Ⓒ The two solutions are equal.

   Ⓓ The relationship cannot be determined from the information given.

|      | Column A | Column B |
|------|----------|----------|
| 9.   | $-8x = 24$ | $\frac{1}{2}y = -4$ |
| 10.  | $-3 = -\frac{1}{4}x$ | $\frac{2}{3}y = 8$ |
| 11.  | $\frac{x}{5} = 3\frac{1}{4}$ | $-\frac{5}{6} = -\frac{1}{3}y$ |

NAME _____          DATE _____

# *Standardized Test Practice*

For use with pages 144–149

TEST TAKING STRATEGY   **Think positively during a test. This will help keep up your confidence and enable you to focus on each question.**

1. *Multiple Choice*   Which one of these steps should you use first to solve $5x - 2 = 15$?

   **I.** Subtract 2.          **II.** Add 2.

   **III.** Divide by 5.       **IV.** Multiply by 5.

   Ⓐ  I                        Ⓑ  II

   Ⓒ  III                      Ⓓ  II or III

   Ⓔ  I or III

2. *Multiple Choice*   Solve $-21 = 2x + 15$.

   Ⓐ  $-8$       Ⓑ  $-3$       Ⓒ  $-18$

   Ⓓ  $18$       Ⓔ  $-\frac{4}{3}$

3. *Multiple Choice*   Solve $6x - 5 - 3x = 4$.

   Ⓐ  $3$        Ⓑ  $-3$       Ⓒ  $1$

   Ⓓ  $-1$       Ⓔ  $\frac{1}{3}$

4. *Multiple Choice*   Solve $-4x - 6 - 3x = 5$.

   Ⓐ  $-\frac{11}{7}$   Ⓑ  $-3$   Ⓒ  $1\frac{6}{11}$

   Ⓓ  $3$        Ⓔ  $2\frac{2}{11}$

5. *Multiple Choice*   Solve the equation $8(2x - 1) - 5x = 25$.

   Ⓐ  $-11$      Ⓑ  $-3$       Ⓒ  $1\frac{6}{11}$

   Ⓓ  $3$        Ⓔ  $2\frac{2}{11}$

6. *Multiple Choice*   Solve $-\frac{2}{3}(x + 2) = 12$.

   Ⓐ  $-10$      Ⓑ  $-6$       Ⓒ  $-16$

   Ⓓ  $10\frac{2}{3}$   Ⓔ  $-20$

7. *Multiple Choice*   Solve the equation $32x - 4(7x + 3) = 16$.

   Ⓐ  $\frac{17}{29}$   Ⓑ  $-7$   Ⓒ  $1$

   Ⓓ  $7$        Ⓔ  $-1$

8. *Multiple Choice*   You live in the United States close to the Canadian border. A Canadian meteorologist reports your temperature as being 25°C. Use the formula $F = \frac{9}{5}C + 32$ to find the temperature in degrees Fahrenheit.

   Ⓐ  46°F       Ⓑ  77°F       Ⓒ  37°F

   Ⓓ  102°F      Ⓔ  68°F

9. *Multiple Choice*   You billed your neighbor $52 for work you did around his house. Materials cost you $16, and you charged $4 an hour. How many hours did you work?

   Ⓐ  7 hours              Ⓑ  8 hours

   Ⓒ  9 hours              Ⓓ  17 hours

   Ⓔ  32 hours

*Quantitative Comparison*   In Exercises 10–13, choose the statement that is the true about the given numbers.

   Ⓐ  The solution in column A is greater.

   Ⓑ  The solution in column B is greater.

   Ⓒ  The two solutions are equal.

   Ⓓ  The relationship cannot be determined from the information given.

| | Column A | Column B |
|---|---|---|
| 10. | $2x - 7 = -15$ | $-3y + 6 = 18$ |
| 11. | $3x - 2(x + 4) = 8$ | $\frac{2}{5}(y - 4) = 6$ |
| 12. | $14(2 - x) = 35$ | $12y + 7 - 5y = -14$ |
| 13. | $7x + 2x = -72$ | $\frac{1}{3}(x - 1) = -3$ |

NAME _____   DATE _____

# *Standardized Test Practice*

For use with pages 151–156

**TEST TAKING STRATEGY   Be aware of how much time you have left, but keep focused on your work.**

1. *Multiple Choice*   Solve $3x - 6 = 4x + 8$.
   - (A) $-2$
   - (B) $2$
   - (C) $-14$
   - (D) $14$
   - (E) $-1$

2. *Multiple Choice*   Solve the equation $5x - 12 = -3x + 6$.
   - (A) $-2\frac{1}{4}$
   - (B) $2\frac{1}{4}$
   - (C) $\frac{3}{4}$
   - (D) $-\frac{3}{4}$
   - (E) $9$

3. *Multiple Choice*   Solve the equation $4x - 2 + 6 = 12 - 5x$.
   - (A) $\frac{8}{9}$
   - (B) $1\frac{1}{8}$
   - (C) $-\frac{8}{9}$
   - (D) $1\frac{1}{9}$
   - (E) $14$

4. *Multiple Choice*   Solve $x + 3 = x - 6$.
   - (A) $3$
   - (B) $6$
   - (C) $-3$
   - (D) $-6$
   - (E) No solution

5. *Multiple Choice*   Which equations are equivalent?
   - I. $2a - 6 = 5 + 9a$
   - II. $5a = 11 + 12a$      III. $-11 = 7a$
   - (A) I and II
   - (B) I and III
   - (C) II and III
   - (D) All
   - (E) None of these

6. *Multiple Choice*   For which equation is $m = 7$ a solution?
   - (A) $m - 5 = 3m - 18$
   - (B) $m - 5 = 3m - 19$
   - (C) $2m = 3m - 17$
   - (D) $2m = -3m - 30$
   - (E) $2m + 1 = -3m - 30$

7. *Multiple Choice*   A local health club has a large swimming pool. Non-members must pay a $6 per day fee to use the pool. Members of the club pay $1, but must pay a yearly membership fee of $250. After how many days using the pool can you justify joining the club? Solve the equation $6x = x + 250$, where $x$ is the number of days you use the pool.
   - (A) 36 days
   - (B) 38 days
   - (C) 50 days
   - (D) 40 days
   - (E) 45 days

8. *Multiple Choice*   Your club decides to sell boxes of stationery to raise money. Each box costs you $2.50 and there is a one time delivery fee of $30. You plan to sell the stationery for $4.50 per box. Which equation should you use to determine how many boxes you must sell to cover your costs?
   - (A) $4.50x + 30 = 2.50x$
   - (B) $2.50x - 30 = 4.50x$
   - (C) $30 - 2.50x = 4.50x$
   - (D) $2.50x + 30 = 4.50x$
   - (E) $30 - 4.50x = 2.50x$

*Quantitative Comparison*   In Exercises 9–11, choose the statement that is the true about the given numbers.
   - (A) The solution in column A is greater.
   - (B) The solution in column B is greater.
   - (C) The two solutions are equal.
   - (D) The relationship cannot be determined from the information given.

| | Column A | Column B |
|---|---|---|
| 9. | $4x + 8 = 2(2x + 4)$ | $3x = 7x - 6$ |
| 10. | $16 - 2x = 2x$ | $\frac{1}{2}(x - 4) = 4$ |
| 11. | $6x - 1 = 5x + 1$ | $4x - 6 = 7x - 12$ |

**Algebra 1, Concepts and Skills**
Standardized Test Practice Workbook

Chapter 3

NAME _____ DATE _____

# *Standardized Test Practice*

**For use with pages 157–162**

TEST TAKING STRATEGY    **Spend no more than a few minutes on each question.**

1. *Multiple Choice*   Which inverse operation can be used to solve the equation $5x = 13$ ?

   (A)  Add 5 to each side.

   (B)  Subtract 5 from each side.

   (C)  Multiply each side by 5.

   (D)  Divide each side by 5.

   (E)  None of the above

2. *Multiple Choice*   Which inverse operation can be used to solve the equation $4 = y - 8$?

   (A)  Add 8 to each side.

   (B)  Subtract 8 from each side.

   (C)  Multiply each side by 8.

   (D)  Divide each side by 8.

   (E)  None of these

3. *Multiple Choice*   Solve $x - 4 = 3(1 - x)$.

   (A)  3          (B)  $\frac{7}{2}$

   (C)  $\frac{3}{2}$          (D)  $\frac{1}{4}$

   (E)  $\frac{7}{4}$

4. *Multiple Choice*   Solve $\frac{1}{2}(6y - 3) = 2 - y$.

   (A)  $\frac{3}{4}$          (B)  $\frac{7}{8}$

   (C)  1          (D)  $\frac{8}{7}$

   (E)  $\frac{3}{2}$

5. *Multiple Choice*   Solve the equation $4x - 3(x + 2) = 5(4 - x)$.

   (A)  $-4\frac{1}{3}$          (B)  $4\frac{1}{3}$          (C)  $\frac{3}{13}$

   (D)  $-\frac{3}{13}$          (E)  No solution

6. *Multiple Choice*   Solve the equation $-10 + 2x + 3(5 - x) = \frac{1}{2}(4x - 8)$.

   (A)  $-3$          (B)  3          (C)  $\frac{1}{3}$

   (D)  $-\frac{1}{3}$          (E)  13

7. *Multi-Step Problem*   A swimming pool club charges nonmembers $10 per day to use the pool. Members pay a yearly fee of $300 and $5 per day to use the pool.

   a.  You want to find out after how many times using the pool you can justify becoming a member. Which equation represents this situation?

   (A)  $10x = (300 + 5)x$

   (B)  $10x + 300 = 5x$

   (C)  $10x = 300 + 5x$

   (D)  $10x = (300 + 5)x$

   (E)  $(10 + 300)x = 5x$

   b.  Solve the equation.

   c.  Explain what the solution means.

NAME _____ DATE _____

# Standardized Test Practice

**For use with pages 163–169**

**TEST TAKING STRATEGY** **Read all of the choices before deciding which is the correct one.**

1. **Multiple Choice** Evaluate 4.36(7.32). Round to the nearest tenth.
   - (A) 31.91
   - (B) 31.92
   - (C) 31.9
   - (D) 32.0
   - (E) 31.0

2. **Multiple Choice** Solve $17x - 19 = 106$. Round to the nearest hundredth.
   - (A) 5.12
   - (B) 5.118
   - (C) 7.36
   - (D) 5.11
   - (E) 7.35

3. **Multiple Choice** Solve $-36 - 19x = 14$. Round to the nearest tenth.
   - (A) 2.6
   - (B) 2.7
   - (C) 1.16
   - (D) 1.2
   - (E) −2.6

4. **Multiple Choice** Solve the equation $14x - 8 = -33 + 2x$. Round to the nearest hundredth.
   - (A) 2.08
   - (B) −3.42
   - (C) −1.56
   - (D) −2.08
   - (E) −2.09

5. **Multiple Choice** Solve the equation $-12.6 + 10.5x = 8x - 7.2$. Round to the nearest tenth.
   - (A) 1.1
   - (B) 2.2
   - (C) −0.3
   - (D) 2.1
   - (E) 1.0

6. **Multiple Choice** Solve the equation $-0.62x - 0.4 = -2.2x + 3.3$. Round to the nearest tenth.
   - (A) 2.3
   - (B) −2.3
   - (C) 1.84
   - (D) 1.8
   - (E) 2.34

7. **Multiple Choice** Solve the equation $-3.68x = -20.52x - 1.71$. Round to the nearest tenth.
   - (A) −9.8
   - (B) −0.1
   - (C) −14.15
   - (D) 0.1
   - (E) −0.2

8. **Multiple Choice** You are shopping for a birthday present. You have $20.00 to spend and your state's sales tax is 6%. Which algebraic model could be used to determine your price limit for the present?
   - (A) $0.06x = 20$
   - (B) $1.06x = 20$
   - (C) $x = 20(0.06)$
   - (D) $x = 20(1.06)$
   - (E) None of these

9. **Multiple Choice** Determine the maximum amount of money you can spend for the present in Exercise 8.
   - (A) $12.00
   - (B) $18.87
   - (C) $18.86
   - (D) $18.80
   - (E) $21.20

10. **Multi-Step Problem** Your cellular phone company charges $25 per month and gives you 30 minutes of free calls per month. Additional minutes cost $.15 per minute. Your bill last month came to $32.80. Let $x$ represent the number of minutes you talked last month.

    a. Which equation models the situation?
       - (A) $25 + 0.15x - 30 = 32.80$
       - (B) $25 + 0.15(x - 30) = 32.80$

    b. For how many minutes did you use the phone?

    c. For how many minutes did you have to pay extra money?

**Algebra 1, Concepts and Skills**
Standardized Test Practice Workbook

Chapter 3

NAME _____  DATE _____

# Standardized Test Practice

**For use with pages 171–176**

**TEST TAKING STRATEGY**  Learn as much as you can about a test ahead of time, such as the types of questions and the topics that the test will cover.

1. **Multiple Choice**  Using the formula for interest, $I = Prt$, solve for $r$.

   **A** $r = \dfrac{Pt}{I}$  **B** $r = \dfrac{I}{Pt}$

   **C** $r = I - Pt$  **D** $r = I - P - t$

   **E** $r = IPt$

2. **Multiple Choice**  Which formulas correctly show the relationship between the perimeter, length, and width of a rectangle?

   **I.** $P = 2l + 2w$

   **II.** $P = 2(l + w)$

   **III.** $w = \dfrac{P}{2} - 2l$

   **A** I only  **B** II only

   **C** III only  **D** I and II only

   **E** I and III

3. **Multiple Choice**  Find the height of a triangle if the base measures 6 inches and the area is 15 square inches. Use the formula for the area of a triangle $A = \dfrac{1}{2}bh$.

   **A** 3 in.  **B** 4 in.

   **C** 5 in.  **D** 6 in.

   **E** 7 in.

4. **Multiple Choice**  Solve the equation $3x + 2y = 10$ for $y$.

   **A** $y = -3x + 5$  **B** $x = -\dfrac{2y}{3} + \dfrac{10}{3}$

   **C** $y = -\dfrac{3}{2}x + 5$  **D** $x = -2y + 7$

   **E** $2y = 10 - 3x$

5. **Multiple Choice**  What is the equivalent of 5°C in degrees Fahrenheit? Use the formula $C = \dfrac{5}{9}(F - 32)$.

   **A** 29.2°F  **B** $-15$°F

   **C** $-20.6$°F  **D** 41°F

   **E** 23°F

6. **Multiple Choice**  Solve $r - s = t$ for $s$.

   **A** $r = s + t$  **B** $s = t - r$

   **C** $s = r + t$  **D** $s = r - t$

   **E** $s = rt$

**Quantitative Comparison**  In Exercises 7–9, choose the statement that is true about the given numbers.

   **A**  The quantity in column A is greater.

   **B**  The quantity in column B is greater.

   **C**  The two quantities are equal.

   **D**  The relationship can not be determined from the information given.

| | Column A | Column B |
|---|---|---|
| 7. | The value of $r$ in the formula $d = rt$ when $d = 25$ and $t = 6$. | The value of $r$ in the formula $d = rt$ when $d = 36$ and $t = 8$. |
| 8. | The value of $y$ in the function $y = 2x - 10$ when $x = 5$. | The value of $y$ in the function $y = \frac{1}{2}x - 8$ when $x = 16$. |
| 9. | The value of $y$ in the function $y = \frac{2}{3}x + 12$ when $x = 9$. | The value of $y$ in the equation $2y - x = 3y - 12$ when $x = 2$. |

# Standardized Test Practice

**For use with pages 177–182**

**TEST TAKING STRATEGY**   **Before you give up on a question, try to eliminate some of your choices so you can make an educated guess.**

**1.** *Multiple Choice*   Which would be the correct unit rate for $3 for 12 cards?

  Ⓐ   $3 per card    Ⓑ   3 cards per dollar

  Ⓒ   $.25 per dozen  Ⓓ   $.25 per card

  Ⓔ   None of these

**2.** *Multiple Choice*   Which model would you use to change 12 minutes to seconds?

  Ⓐ   $12 \text{ min} \cdot \dfrac{1 \text{ min}}{60 \text{ sec}}$  Ⓑ   $12 \text{ min} \cdot \dfrac{60 \text{ sec}}{1 \text{ min}}$

  Ⓒ   $12 \text{ min} \cdot \dfrac{60 \text{ min}}{1 \text{ sec}}$  Ⓓ   $12 \text{ min} \cdot \dfrac{1 \text{ sec}}{60 \text{ min}}$

  Ⓔ   None of these

**3.** *Multiple Choice*   Find the average speed of a person boating 126 miles in 5 hours.

  Ⓐ   25.2 miles    Ⓑ   25.2 mi/h

  Ⓒ   630 mi/h    Ⓓ   3.9 mi/h

  Ⓔ   3.9 hours per mile

**4.** *Multiple Choice*   Convert 250 pesos to dollars. The exchange rate is 9.99 pesos per United States dollar.

  Ⓐ   $2497.50    Ⓑ   $259.99

  Ⓒ   $25.03    Ⓓ   $240.01

  Ⓔ   $30.02

**5.** *Multiple Choice*   Which model would you use to convert 35 miles per hour to feet per second?

  Ⓐ   $35 \text{ mi/h} \times \dfrac{5280 \text{ ft}}{1 \text{ mile}} \times \dfrac{60 \text{ min}}{1 \text{ h}}$

  Ⓑ   $35 \text{ mi/h} \times \dfrac{1 \text{ mile}}{5280 \text{ ft}} \times \dfrac{1 \text{ h}}{60 \text{ min}}$

  Ⓒ   $35 \text{ mi/h} \times \dfrac{5280 \text{ ft}}{1 \text{ mile}} \times \dfrac{1 \text{ h}}{360 \text{ sec}}$

  Ⓓ   $35 \text{ mi/h} \times \dfrac{5280 \text{ ft}}{1 \text{ mile}} \times \dfrac{1 \text{ h}}{3600 \text{ sec}}$

  Ⓔ   $35 \text{ mi/h} \times \dfrac{5280 \text{ ft}}{1 \text{ mile}} \times \dfrac{1 \text{ h}}{60 \text{ min}}$

**6.** *Multiple Choice*   Write the ratio 128 to 24 in simplest form.

  Ⓐ   5    Ⓑ   $\dfrac{64}{12}$

  Ⓒ   $\dfrac{32}{6}$    Ⓓ   $\dfrac{16}{3}$

  Ⓔ   $\dfrac{15}{3}$

**7.** *Multiple Choice*   A kennel has 15 cats and 24 dogs. What is the ratio in simplest form of dogs to cats?

  Ⓐ   $\dfrac{8}{5}$    Ⓑ   $\dfrac{24}{15}$

  Ⓒ   $\dfrac{6}{3}$    Ⓓ   $\dfrac{5}{3}$

  Ⓔ   $\dfrac{3}{2}$

**8.** *Multiple Choice*   What does the unit conversion factor $\dfrac{12 \text{ inches}}{1 \text{ foot}}$ equal?

  Ⓐ   12    Ⓑ   $\dfrac{1}{12}$

  Ⓒ   1    Ⓓ   cannot be determined

  Ⓔ   None of these

**9.** *Multi-Step Problem*   You record the amount of gas and the cost when you filled the tank in your car.

| Number of gallons | 12.2 | 15.3 | 14.4 |
|---|---|---|---|
| Cost (dollars) | 18.09 | 23.35 | 23.35 |

  **a.** Describe how to compute the average cost of gas.

  **b.** Find the average cost of gas. Round your answer to the nearest $.01.

  **c.** Use your answer from part (b) to estimate the cost of filling an empty tank if your car's tank holds 16 gallons.

Chapter 3

NAME _____     DATE _____

# Standardized Test Practice

**For use with pages 183–188**

**TEST TAKING STRATEGY**   Avoid spending too much time on one question. Skip questions that are too difficult for you and spend no more than a few minutes on each question.

1. *Multiple Choice*   What is 12% of 136?

   Ⓐ  11.33       Ⓑ  16.32       Ⓒ  15.21

   Ⓓ  14.7        Ⓔ  27.1

2. *Multiple Choice*   What is 420% of 48?

   Ⓐ  87.5        Ⓑ  20.16       Ⓒ  100

   Ⓓ  201.6       Ⓔ  875

3. *Multiple Choice*   13 is what percent of 57?

   Ⓐ  about 22.8%   Ⓑ  about 7.41%

   Ⓒ  about 4.38%   Ⓓ  about 438.5%

   Ⓔ  about 741%

4. *Multiple Choice*   75 is what percent of 20?

   Ⓐ  3.75%          Ⓑ  about 26.7%

   Ⓒ  15%            Ⓓ  about 267%

   Ⓔ  375%

5. *Multiple Choice*   31.5 is 15% of what number?

   Ⓐ  200         Ⓑ  210         Ⓒ  230

   Ⓓ  473         Ⓔ  476

6. *Multiple Choice*   90 is 120% of what number?

   Ⓐ  108         Ⓑ  133         Ⓒ  75

   Ⓓ  62          Ⓔ  85

7. *Multiple Choice*   A store has a 50% off sale in September on select items. In October, it marks the sale prices back up by 50%. What would a sweater that originally sold for $100 now be selling for?

   Ⓐ  $100        Ⓑ  $50         Ⓒ  $150

   Ⓓ  $25         Ⓔ  $75

*Multiple Choice*   In Exercises 8 and 9, refer to the graph. In a survey, people were asked if they think the income tax system is fair to everyone.

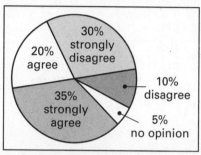

8. In the survey, 91 people strongly agreed that the income tax system is fair. How many people were surveyed?

   Ⓐ  250         Ⓑ  3185        Ⓒ  319

   Ⓓ  455         Ⓔ  260

9. If you asked 550 people, how many would you expect to answer "disagree" or "strongly disagree"?

   Ⓐ  165         Ⓑ  55          Ⓒ  220

   Ⓓ  250         Ⓔ  303

*Quantitative Comparison*   In Exercises 10–13, find the value of $x$. Then choose the statement below that is true about the given numbers.

   Ⓐ  The value in column A is greater.

   Ⓑ  The value in column B is greater.

   Ⓒ  The two values are equal.

   Ⓓ  The relationship cannot be determined from the given information.

|     | Column A | Column B |
| --- | --- | --- |
| 10. | 57% of 200 = $x$ | 110% of 60 = $x$ |
| 11. | 32% of $x$ = 540 | 16% of $x$ = 270 |
| 12. | $x$ = 17% of 28 | 25% of $x$ = 1.2 |
| 13. | $x$% of 72 = 18 | 20% of 125 = $x$ |

NAME _____   DATE _____

# *Standardized Test Practice*

**For use with pages 203–208**

**TEST TAKING STRATEGY**   **Read all of the answer choices before deciding which is the correct one.**

*Multiple Choice*   For Exercises 1–5, refer to the coordinate plane below.

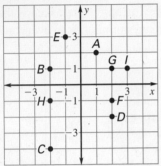

1. Which quadrant is point *B* in?

    Ⓐ   Quadrant I       Ⓑ   Quadrant II

    Ⓒ   Quadrant III     Ⓓ   Quadrant IV

    Ⓔ   Cannot be determined from the given information

2. Which ordered pair corresponds to the point labeled *A*?

    Ⓐ   (1, 2)        Ⓑ   (2, 1)

    Ⓒ   (−1, 2)       Ⓓ   (−1, −2)

    Ⓔ   (2, −1)

3. Which point corresponds to the ordered pair (2, −1)?

    Ⓐ   *G*          Ⓑ   *B*          Ⓒ   *H*

    Ⓓ   *D*          Ⓔ   *F*

4. Which quadrant would the ordered pair (−3, −2) be in?

    Ⓐ   I           Ⓑ   II          Ⓒ   III

    Ⓓ   IV          Ⓔ   II or III

5. Which points have positive *y*-coordinates?

    Ⓐ   *A* and *F*      Ⓑ   *A* and *B*

    Ⓒ   *B* and *H*      Ⓓ   *B* and *F*

    Ⓔ   *H* and *F*

6. *Multiple Choice*   Use the scatter plot below. Which of the following is true?

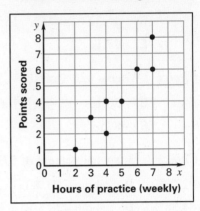

**Hours of practice (weekly)**

    Ⓐ   Points scored decreases as hours of practice increases.

    Ⓑ   Points scored increases as hours of practice decreases.

    Ⓒ   Points scored increases as hours of practice increases.

    Ⓓ   Points scored is not related to hours of practice increases.

    Ⓔ   None of these

7. *Multi-Step Problem*   The number of skiers (in thousands) at a winter resort are shown in the table.

| Year | 1992 | 1993 | 1994 |
|------|------|------|------|
| Skiers | 8.2 | 8.6 | 9.1 |

| Year | 1995 | 1996 | 1997 |
|------|------|------|------|
| Skiers | 9.7 | 10.3 | 10.8 |

   a. Construct a scatter plot of the data. Use the horizontal axis to represent the number of years since 1992.

   b. Describe the relationship between the number of skiers and the number of years since 1992.

**TEST TAKING STRATEGY**   **Learn as much as you can about a test ahead of time, such as the types of questions and the topics that the test will cover.**

1. **Multiple Choice**   Choose the ordered pair that is a solution of $2x - 3y = 6$.

   **A**   $(5, 1)$      **B**   $(0, 6)$      **C**   $(4, 3)$

   **D**   $(6, 2)$      **E**   $(3, 2)$

2. **Multiple Choice**   Choose the ordered pair that is a solution of $y = 2x - 1$.

   **A**   $(1, 3)$      **B**   $(-1, -3)$

   **C**   $(-3, 0)$      **D**   $(3, -1)$

   **E**   $(-3, -1)$

3. **Multiple Choice**   Choose the group of ordered pairs that are solutions of the equation $y = -3x + 4$.

   **A**   $(1, 0), (2, -2)$   **B**   $(1, 2), (3, 4)$

   **C**   $(1, 1), (3, -5)$   **D**   $(0, 4), (2, 2)$

   **E**   $(2, -3), (4, 5)$

4. **Multiple Choice**   Choose the group of ordered pairs that are solutions of the equation $y = 2x - 16$.

   **A**   $(0, -16), (2, -12)$

   **B**   $(0, -14), (1, -12)$

   **C**   $(1, -16), (4, -8)$

   **D**   $(4, -8), (8, 8)$

   **E**   $(8, 8), (2, -12)$

5. **Multiple Choice**   Rewite $3x - y = -6$ in function form.

   **A**   $3x = y - 6$   **B**   $x = \frac{1}{3}y + 2$

   **C**   $y = 3x + 6$   **D**   $3x = y + 6$

   **E**   $y = 3x - 6$

6. **Multiple Choice**   Which point does *not* lie on the graph of $3y + \frac{1}{2}x = 8$?

   **A**   $\left(2, \frac{7}{3}\right)$      **B**   $\left(0, \frac{8}{3}\right)$   **C**   $(1, 2)$

   **D**   $(4, 2)$      **E**   $(10, 1)$

7. **Multiple Choice**   Which equation does the graph represent?

   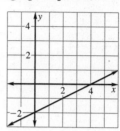

   **A**   $x - 2y = 4$      **B**   $x + 2y = 4$

   **C**   $-x + 2y = 4$   **D**   $-x - 2y = 4$

   **E**   $-x - 2y = -4$

8. **Multiple Choice**   Which equation does the graph represent?

   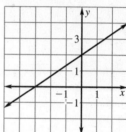

   **A**   $2x + 3y = 6$      **B**   $2x + 3y = -6$

   **C**   $-2x + 3y = 6$   **D**   $-2x - 3y = 6$

   **E**   $2x - 3y = 6$

9. **Multi-Step Problem**   Your parents limit your phone use to 18 hours every week. You need to split that between hours spent on the phone with your friends and hours spent using the Internet. Your parents count one hour on line as $\frac{1}{2}$ hour of phone time since some of your time is spent doing homework. An algebraic model is $x + \frac{1}{2}y = 18$, where $x$ is hours spent on the phone and $y$ is the hours spent on line.

   **a.** Solve the equation for $y$.

   **b.** Use the equation from part (a) to make a table of values for $x = 5$, $x = 10$, and $x = 15$.

   **c.** Plot the points and draw the line.

**Algebra 1, Concepts and Skills**
Standardized Test Practice Workbook

**27**

*Chapter 4*

NAME _____ DATE _____

# Standardized Test Practice

For use with pages 216–221

**TEST TAKING STRATEGY**   **Spend no more than a few minutes on each question.**

1. *Multiple Choice*   Which point does not lie on the graph of $x = -4$?

   Ⓐ $(-4, 0)$    Ⓑ $(-4, 1)$

   Ⓒ $(4, -4)$    Ⓓ $(-4, -4)$

   Ⓔ $(-4, 2)$

2. *Multiple Choice*   The ordered pair $(-1, 4)$ is a solution of _____ .

   Ⓐ $x = 4$    Ⓑ $x = 1$

   Ⓒ $y = 1$    Ⓓ $y = -1$

   Ⓔ $y = 4$

3. *Multiple Choice*   Which of the following ordered pairs is a solution of $x = 2$?

   Ⓐ $(1, 3)$    Ⓑ $(2, 0)$

   Ⓒ $(0, 1)$    Ⓓ $(1, 0)$

   Ⓔ $(1, 2)$

4. *Multiple Choice*   Which of the following ordered pairs is not solution of $y = -1$?

   Ⓐ $(-1. -1)$    Ⓑ $(0. -1)$

   Ⓒ $\left(-\frac{1}{2}, -1\right)$    Ⓓ $(-1, 1)$

   Ⓔ $(2, -1)$

5. *Multiple Choice*   The ordered pair $(-2, -1)$ is a solution of _____ .

   Ⓐ $x = -1$    Ⓑ $x = 1$

   Ⓒ $x = 0$    Ⓓ $y = 1$

   Ⓔ $y = -1$

6. *Multiple Choice*   Which equation is equivalent to $y = 2$?

   Ⓐ $y = 0 \cdot x + 2$    Ⓑ $0 \cdot y = 2$

   Ⓒ $0 \cdot y = 1 \cdot x + 2$    Ⓓ $y = 2 \cdot 0 + x$

   Ⓔ $y = x + 2$

7. *Multiple Choice*   What is the equation of the line shown?

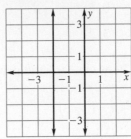

   Ⓐ $x = -2$    Ⓑ $x = 2$

   Ⓒ $y = -2$    Ⓓ $y = 2$

   Ⓔ $y = 0$

8. *Multi-Step Problem*

   a. Why is the function $y = 3$ called a constant function?

   b. What is the domain of the function $y = 3$?

   c. What is the range of the function $y = 3$?

   d. Name the point that lies on the graph of $y = 3$ and also on the graph of $x = -4$.

**Algebra 1, Concepts and Skills**
Standardized Test Practice Workbook

NAME _____ DATE _____

# *Standardized Test Practice*

For use with pages 222–227

**TEST TAKING STRATEGY** **Some questions involve more than one step. Reading too quickly might lead to mistaking the answer to a preliminary step for your final answer.**

1. *Multiple Choice* What is the $x$-intercept of the line $5x - 2y = 10$?

   (A) 0      (B) 2      (C) 5

   (D) $-2$      (E) $-5$

2. *Multiple Choice* What is the $y$-intercept of the line $5x - 2y = 10$?

   (A) 0      (B) 2      (C) 5

   (D) $-2$      (E) $-5$

3. *Multiple Choice* What is the $x$-intercept of the line $\frac{1}{2}x + \frac{3}{5}y = 20$?

   (A) 0      (B) 12      (C) $\frac{100}{3}$

   (D) 10      (E) 40

4. *Multiple Choice* What is the $y$-intercept of the line shown?

   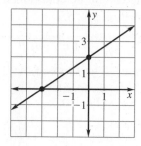

   (A) 3      (B) $-3$      (C) 2

   (D) $-2$      (E) 0

5. *Multiple Choice* What is the $x$-intercept of the line shown in Exercise 4?

   (A) 3      (B) $-3$      (C) 2

   (D) $-2$      (E) 0

6. *Multiple Choice* A hot dog stand sells foot-long hot dogs for $2.50 and 6 inch hot dogs for $1.50. If the stand makes $54 one day, which of the following model(s) the situation?

   (A) $2.50x - 1.50y = 54$

   (B) $2.50x + 1.50y = 54$

   (C) $y = -\frac{5}{3}x - 36$

   (D) $y = -\frac{5}{3}x + 36$

   (E) B and D

7. What does the y-intercept of the graph of the equation in Exercise 6 represent?

   (A) The amount of money earned from foot-long hot dogs

   (B) The amount of money earned from 6 inch hot dogs

   (C) The number of foot-long hot dogs sold if no 6 inch hot dogs are sold

   (D) The number of 6 inch hot dogs sold if no foot-long hot dogs are sold

   (E) The cost of a 6 inch hot dog

*Quantitative Comparison* In Exercises 8–10, choose the statement that is true about the given numbers.

   (A) The number in column A is greater.

   (B) The number in column B is greater.

   (C) The two numbers are equal.

   (D) The relationship cannot be determined from the information given.

| | Column A | Column B |
|---|---|---|
| 8. | The $x$-intercept of $3y + 6x = 12$ | The $y$-intercept of $2y + 7x = 14$ |
| 9. | 0 | The $y$-intercept of $x + \frac{1}{2}y = 4$ |
| 10. | The $x$-intercept of $\frac{1}{2}x - 2y = 8$ | The $y$-intercept of $\frac{1}{2}x - 2y = 8$ |

Chapter 4

# Standardized Test Practice

**For use with pages 229–235**

**TEST TAKING STRATEGY    Spend no more than a few minutes on each question.**

**1. Multiple Choice**   Which is the formula to find the slope of a line?

(A)  $m = \dfrac{x_2 - x_1}{y_2 - y_1}$      (B)  $m = \dfrac{y_2 + y_1}{x_2 + x_1}$

(C)  $m = \dfrac{y_2 - y_1}{x_2 - x_1}$      (D)  $m = \dfrac{x_2 + x_1}{y_2 - y_1}$

(E)  $m = \dfrac{y_2 - x_2}{y_1 - x_1}$

**2. Multiple Choice**   Find the slope of the line passing through the points $(5, 8)$ and $(9, 4)$.

(A)  3            (B)  1            (C)  $\frac{3}{4}$

(D)  $-1$         (E)  $\frac{4}{3}$

**3. Multiple Choice**   Find the slope of the line passing through the points $(-6, 7)$ and $(2, 9)$.

(A)  $\frac{1}{4}$      (B)  $-2$      (C)  4

(D)  $-4$              (E)  $-\frac{1}{2}$

**4. Multiple Choice**   Find the slope of the line passing through the points $(3, 4)$ and $(-2, 4)$.

(A)  undefined    (B)  0            (C)  8

(D)  $\frac{8}{5}$        (E)  $\frac{5}{8}$

**5. Multiple Choice**   Find the slope of the line passing through the points $(7, -2)$ and $(7, 9)$.

(A)  undefined    (B)  0            (C)  $\frac{11}{14}$

(D)  $\frac{1}{2}$        (E)  2

**6. Multiple Choice**   Find the value of $y$ so that the line passing through $(2, 6)$ and $(1, y)$ has a slope of 5.

(A)  2            (B)  9            (C)  $-2$

(D)  $-1$         (E)  1

**7. Multiple Choice**   Find the value of $x$ so that the line passing through $(10, 5)$ and $(x, 9)$ has a slope of $-2$.

(A)  2        (B)  $-2$        (C)  8

(D)  $-8$         (E)  $-17$

**8. Multiple Choice**   The slope of a horizontal line is ___?___.

(A)  undefined    (B)  zero

(C)  positive     (D)  negative

(E)  variable

**Quantitative Comparison**   In Exercises 9–12, choose the statement that is true about the given numbers.

(A)  The number in column A is greater.

(B)  The number in column B is greater.

(C)  The two numbers are equal.

(D)  The relationship cannot be determined from the information given.

|     | Column A | Column B |
|-----|----------|----------|
| 9.  | The slope of the line through $(6, 6)$ and $(10, 8)$ | The slope of the line through $(5, 8)$ and $(17, 14)$ |
| 10. | The slope of the line through $(3, 6)$ and $(5, 3)$ | 0 |
| 11. | The slope of the line through $(1, 3)$ and $(8, 10)$ | The slope of the line through $(12, 6)$ and $(18, 0)$ |
| 12. | The slope of the line through $(5, y)$ and $(7, 3)$ | The slope of the line through $(4, 8)$ and $(x, 14)$ |

# *Standardized Test Practice*

**For use with pages 236–241**

**TEST TAKING STRATEGY**    **Go back and check as much of your work as you can.**

1. *Multiple Choice*   In a direct variation model, if the constant of variation in the equation is 3, then the slope is ___?___.

   Ⓐ 3          Ⓑ $-3$

   Ⓒ $\frac{1}{3}$          Ⓓ $-\frac{1}{3}$

   Ⓔ Cannot be determined from the given information

2. *Multiple Choice*   Choose the constant of variation of the direct variation model $y = -5x$.

   Ⓐ 5          Ⓑ $-5$          Ⓒ $-\frac{1}{5}$

   Ⓓ $x$          Ⓔ $y$

3. *Multiple Choice*   Choose the constant of variation of the direct variation model $-\frac{1}{2}x = y$.

   Ⓐ $x$          Ⓑ $-2$          Ⓒ $\frac{1}{2}$

   Ⓓ $-\frac{1}{2}$          Ⓔ $y$

4. *Multiple Choice*   The variables $x$ and $y$ vary directly. Which equation relates $x$ and $y$ when $x = 4$ and $y = 12$?

   Ⓐ $x = 4y$          Ⓑ $y = 4x$

   Ⓒ $x = 3y$          Ⓓ $y = 12x$

   Ⓔ $y = 3x$

5. *Multiple Choice*   The variables $x$ and $y$ vary directly. Which equation relates $x$ and $y$ when $x = \frac{1}{2}$ and $y = 6$?

   Ⓐ $x = 3y$          Ⓑ $y = 3x$

   Ⓒ $x = 12y$          Ⓓ $y = 12x$

   Ⓔ $y = \frac{1}{2}x$

6. *Multiple Choice*   Which property is true for the graph of a direct variation model?

   Ⓐ The slope is always positive.

   Ⓑ The slope is always negative.

   Ⓒ The graph always contains (0,0).

   Ⓓ The slope is always 0.

   Ⓔ The $y$-intercept is always positive.

7. *Multiple Choice*   One year in a dog's life is equivalent to seven years in a human's life. Choose the model that relates a dog's life $d$ to a human's life $h$.

   Ⓐ $d = -7h$   Ⓑ $d = 7h$

   Ⓒ $d = \frac{1}{7}h$   Ⓓ $h = \frac{1}{7}d$

   Ⓔ $h = -\frac{1}{7}d$

*Quantitative Comparison*   In Exercises 8–10, choose the statement that is true about the given numbers.

   Ⓐ The number in column A is greater.

   Ⓑ The number in column B is greater.

   Ⓒ The two numbers are equal.

   Ⓓ The relationship cannot be determined from the information given.

| | Column A | Column B |
|---|---|---|
| 8. | The constant of variation in $y = -3x$ | The constant of variation in $5x = y$ |
| 9. | The value of $y$ when $x = 3$ and $y = -\frac{1}{9}x$ | The value of $x$ when $y = -2$ and $y = 2x$ |
| 10. | The slope of the line with equation $y = -\frac{1}{2}x$ | The constant of variation in $y = -\frac{1}{2}x$ |

**Algebra 1, Concepts and Skills**
Standardized Test Practice Workbook

**TEST TAKING STRATEGY**   **Think positively during a test. This will help keep up your confidence and enable you to focus on each question.**

1. *Multiple Choice*   What is the slope of a linear equation written in slope-intercept form, $y = mx + b$?

   (A)  $y$          (B)  $x$          (C)  $m$

   (D)  $b$          (E)  $mx$

2. *Multiple Choice*   What is the slope of the graph of $y = -3x + 2$?

   (A)  1          (B)  $-3$          (C)  2

   (D)  3          (E)  $-2$

3. *Multiple Choice*   What is the slope of the graph of $4x + 3y = 15$?

   (A)  4          (B)  $-4$          (C)  $-\frac{4}{3}$

   (D)  $\frac{4}{3}$          (E)  5

4. *Multiple Choice*   What is the $y$-intercept of the graph of $y = 5x - 2$?

   (A)  1          (B)  5          (C)  2

   (D)  $-2$          (E)  $-5$

5. *Multiple Choice*   What is the $y$-intercept of the graph of $3y - 2x = 18$?

   (A)  18          (B)  $\frac{2}{3}$          (C)  6

   (D)  $-6$          (E)  $-\frac{2}{3}$

6. *Multiple Choice*   Choose the set of equations which are parallel lines.

   (A)  $x = 2, y = 6$

   (B)  $y = 3x + 6, y = 2x + 6$

   (C)  $y = -5x + 1, y = 5x + 3$

   (D)  $y = -\frac{1}{2}x + 2, y = -\frac{1}{2}x$

   (E)  $y = 6x + 3, y = \frac{1}{6}x + 3$

7. *Multiple Choice*   What is the equation, in slope-intercept form, of the line shown?

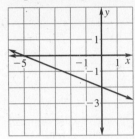

   (A)  $-2x - 5y = 10$  (B)  $y = -\frac{2}{5}x + 2$

   (C)  $y = -\frac{2}{5}x - 2$  (D)  $-5x - 2y = 10$

   (E)  $y = \frac{2}{5}x - 2$

8. *Multiple Choice*   You are filling a pot with water. The water level in the pot is rising at a rate of 2 inches per minute. The pot is already 3 inches full. The equation $y = 2x + 3$ models the depth of the water after $x$ minutes. What is the depth of the water after 3 minutes?

   (A)  3 in.          (B)  7 in.          (C)  6 in.

   (D)  8 in.          (E)  9 in.

*Quantitative Comparison*   In Exercises 9–11, choose the statement that is true about the given numbers.

   (A)  The number in column A is greater.

   (B)  The number in column B is greater.

   (C)  The two numbers are equal.

   (D)  The relationship cannot be determined from the information given.

|     | Column A | Column B |
|-----|----------|----------|
| 9.  | The slope of $y = -3x + 6$ | The slope of $y = -2x + 2$ |
| 10. | The slope of $y = -5x$ | The slope of $2y = -10x + 15$ |
| 11. | The $y$-intercept of $y = \frac{1}{2}x + 3$ | The $y$-intercept of $y = 13x$ |

NAME _____ DATE _____

# *Standardized Test Practice*

For use with pages 252–257

**TEST TAKING STRATEGY**   **Be aware of how much time you have left, but keep focused on your work.**

**1. *Multiple Choice***   Which of the relations below is *not* a function?

Ⓐ
| Input | 1 | 2 | 3 | 4 |
|-------|---|---|---|---|
| Output | 2 | 4 | 6 | 8 |

Ⓑ
| Input | 0 | 1 | 2 | 3 |
|-------|---|---|---|---|
| Output | 2 | 2 | 2 | 2 |

Ⓒ
| Input | 0 | 1 | 1 | 2 |
|-------|---|---|---|---|
| Output | 1 | 2 | 3 | 4 |

Ⓓ
| Input | 1 | 2 | 3 | 4 |
|-------|---|---|---|---|
| Output | 2 | 2 | 4 | 4 |

Ⓔ
| Input | 1 | 2 | 3 | 4 |
|-------|---|---|---|---|
| Output | 5 | 6 | 7 | 8 |

**2. *Multiple Choice***   Evaluate the function $f(x) = \frac{3}{2}x + 6$ when $x = 2$.

Ⓐ 9     Ⓑ 6     Ⓒ $-9$

Ⓓ $-6$     Ⓔ $-\frac{8}{3}$

**3. *Multiple Choice***   Evaluate the function $g(x) = 2x - 7$ for $x = -3$.

Ⓐ $-12$     Ⓑ $-10$     Ⓒ 1

Ⓓ $-13$     Ⓔ $-1$

**4. *Multiple Choice***   Find the slope of the graph of the linear function $f$ if $f(3) = 6$ and $f(-2) = -4$.

Ⓐ $-\frac{2}{5}$     Ⓑ $\frac{1}{2}$     Ⓒ $-2$

Ⓓ 2     Ⓔ $-\frac{5}{2}$

**5. *Multiple Choice***   Find the slope of the graph of the linear function $g$ if $g(-2) = 4$ and $g(2) = 5$.

Ⓐ 0     Ⓑ 9     Ⓒ 4

Ⓓ $-\frac{1}{4}$     Ⓔ $\frac{1}{4}$

**6. *Multiple Choice***   The line below matches which function?

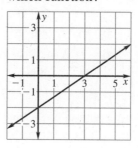

Ⓐ $f(x) = -\frac{2}{3}x - 2$     Ⓑ $f(x) = \frac{2}{3}x - 2$

Ⓒ $f(x) = \frac{2}{3}x + 2$     Ⓓ $f(x) = \frac{2}{3}x + 3$

Ⓔ $f(x) = \frac{2}{3}x - 3$

**7. *Multiple Choice***   If it takes 1.5 hours for a stick to float $\frac{3}{4}$ of a mile downstream, choose the correct linear function expressing the distance traveled as a function of time.

Ⓐ $f(t) = 2t$     Ⓑ $f(t) = 1.5t$

Ⓒ $f(t) = 0.5t$     Ⓓ $f(t) = 0.75t$

Ⓔ $f(t) = 1.125t$

*Quantitative Comparison*   In Exercises 8–10, choose the statement that is true about the given numbers.

Ⓐ   The number in column A is greater.

Ⓑ   The number in column B is greater.

Ⓒ   The two numbers are equal.

Ⓓ   The relationship cannot be determined from the information given.

| | Column A | Column B |
|---|---|---|
| **8.** | $f(x) = -5x + 2$ when $x = 3$ | $f(x) = -5x + 2$ when $x = -3$ |
| **9.** | $g(x) = \frac{1}{2}x - 3$ when $x = 4$ | $g(x) = -\frac{2}{3}x + 3$ when $x = 9$ |
| **10.** | $f(x) = 3x + 3$ when $x = -2$ | $f(x) = 4x + 3$ when $x = -2$ |

**Algebra 1, Concepts and Skills**
Standardized Test Practice Workbook

NAME _____  DATE _____

# *Standardized Test Practice*

**For use with pages 269–275**

**TEST TAKING STRATEGY**   **Be aware of how much time you have left, but keep focused on your work.**

**1.** *Multiple Choice*   An equation of the line whose slope is 3 and whose *y*-intercept is 8 is ___?___ .

**(A)** $y = 8x + 3$    **(B)** $y = 3x + 8$

**(C)** $y = -3x + 8$    **(D)** $y = -3x - 8$

**(E)** $y = 3x - 8$

**2.** *Multiple Choice*   An equation of the line whose slope is $-6$ and whose *y*-intercept is 9 is ___?___ .

**(A)** $y = -9x - 6$    **(B)** $y = 9x - 6$

**(C)** $y = -6x - 9$    **(D)** $y = 6x - 9$

**(E)** $y = -6x + 9$

**3.** *Multiple Choice*   An equation of the line whose slope is $\frac{1}{2}$ and whose *y*-intercept is $-2$ is ___?___ .

**(A)** $y = \frac{1}{2}x - 2$    **(B)** $y = -\frac{1}{2}x - 2$

**(C)** $y = -2x - \frac{1}{2}$    **(D)** $y = -2x + \frac{1}{2}$

**(E)** $y = \frac{1}{2}x + 2$

**4.** *Multiple Choice*   What is an equation of the line shown in the graph?

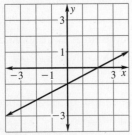

**(A)** $y = 2x - 1$    **(B)** $y = \frac{1}{2}x + 1$

**(C)** $y = \frac{1}{2}x - 1$    **(D)** $y = -\frac{1}{2}x + 1$

**(E)** $y = -x + \frac{1}{2}$

**5.** *Multiple Choice*   Write a set of equations for the parallel lines shown in the graph.

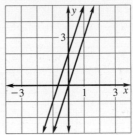

**(A)** $y = 3x; y = 3x - 2$

**(B)** $y = 3x; y = 3x + 2$

**(C)** $y = \frac{1}{3}x; y = \frac{1}{3}x + 2$

**(D)** $y = \frac{1}{3}x; y = \frac{1}{3}x - 1$

**(E)** $y = 3x; y = 3x - 1$

*Quantitative Comparison*   In Exercises 6–9, choose the statement below that is true about the given numbers.

**(A)**   The number is column A is greater.

**(B)**   The number in column B is greater.

**(C)**   The two numbers are equal.

**(D)**   The relationship cannot be determined from the given information.

| | Column A | Column B |
|---|---|---|
| **6.** | The *y*-intercept of $y = 3x - 5$ | The slope of $y = 3x - 5$ |
| **7.** | The slope of $y = -\frac{1}{2}x - \frac{1}{5}$ | The *y*-intercept of $y = -\frac{1}{2}x - \frac{1}{5}$ |
| **8.** | The *x*-intercept of $y = 5x + 6$ | The *y*-intercept of $y = \frac{2}{5}x - \frac{6}{5}$ |
| **9.** | The *y*-intercept of $y = 3x$ | The *x*-intercept of $y = 2x + 1$ |

NAME _____  DATE _____

# Standardized Test Practice

For use with pages 278–284

**TEST TAKING STRATEGY** As soon as the testing begins, start working. Keep a steady pace and stay focused on the test.

1. **Multiple Choice** What is the value of $x_1$ in the equation $y - 6 = \frac{1}{2}(x + 2)$?

   Ⓐ 6     Ⓑ −6     Ⓒ 2

   Ⓓ −2     Ⓔ $\frac{1}{2}$

2. **Multiple Choice** Which equation in point-slope form passes through the point (2, 3) and has a slope of $-\frac{1}{3}$?

   Ⓐ $y + 3 = -\frac{1}{3}(x - 2)$

   Ⓑ $y - 3 = -\frac{1}{3}(x - 2)$

   Ⓒ $y - 2 = -\frac{1}{3}(x - 3)$

   Ⓓ $y + 2 = -\frac{1}{3}(x + 3)$

   Ⓔ $y = -\frac{1}{3}x + \frac{7}{3}$

3. **Multiple Choice** Which equation in point-slope form passes through the point $(-1, 6)$ and has slope of $\frac{2}{3}$?

   Ⓐ $y - 6 = \frac{2}{3}(x + 1)$    Ⓑ $y - 1 = \frac{2}{3}(x + 6)$

   Ⓒ $y + 6 = \frac{2}{3}(x - 1)$    Ⓓ $y - 6 = \frac{2}{3}(x - 1)$

   Ⓔ $y + 1 = \frac{2}{3}(x - 6)$

4. **Multiple Choice** Find an equation of the line that passes through (3, 5) and has slope −3.

   Ⓐ $y = -3x - 4$    Ⓑ $y = 3x + 14$

   Ⓒ $y - 3 = -3(x - 5)$

   Ⓓ $y - 5 = -\frac{1}{3}(x - 3)$

   Ⓔ $y = -3x + 18$

5. **Multiple Choice** Find an equation of the line that is parallel to the line $y = \frac{2}{3}x + 6$ and passes through (6, 2).

   Ⓐ $y - 2 = \frac{2}{3}(x - 6)$

   Ⓑ $y - 6 = \frac{2}{3}(x + 2)$

   Ⓒ $y = \frac{3}{2}x - 11$

   Ⓓ $y = \frac{2}{3}x + 2$

   Ⓔ $y = \frac{2}{3}x - 6$

6. **Multiple Choice** Which is an equation for the line shown in the graph?

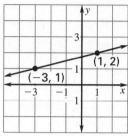

   Ⓐ $y + 3 = -\frac{1}{4}(x - 1)$

   Ⓑ $y - 2 = -4(x - 1)$

   Ⓒ $y - 1 = \frac{1}{4}(x + 3)$

   Ⓓ $-y - 2 = 4(x - 1)$

   Ⓔ $-y - 1 = -\frac{1}{4}(x + 3)$

*Quantitative Comparison* In Exercises 7–9, choose the statement below that is true about the given numbers.

Ⓐ The number is column A is greater.

Ⓑ The number in column B is greater.

Ⓒ The two numbers are equal.

Ⓓ The relationship cannot be determined from the given information.

| | Column A | Column B |
|---|---|---|
| 7. | The value of $m$ in $y + 3 = \frac{1}{2}(x - 2)$ | The value of $x_1$ in $y - 6 = \frac{1}{2}(x - 2)$ |
| 8. | The value of $y_1$ in $y + 3 = -2(x - 3)$ | The value of $m$ in the line passing through (1, 8) and (2, 3) |
| 9. | The value of $m$ in the line passing through $(-1, -5)$ and $(2, -3)$ | The value of $y_1$ in $y - 6 = 7(x - 8)$ |

# *Standardized Test Practice*

**For use with pages 285–290**

**TEST TAKING STRATEGY   Go back and check as much of your work as you can.**

1. *Multiple Choice*   What is an equation of the line that passes through the points $(2, 3)$ and $(-1, 4)$?

   **A**  $y = 3x - \frac{11}{3}$    **B**  $y = -\frac{1}{3}x - \frac{11}{3}$

   **C**  $y = -3x + \frac{11}{3}$    **D**  $y = -\frac{1}{3}x + \frac{11}{3}$

   **E**  $y = -\frac{1}{3}x + 3$

2. *Multiple Choice*   What is an equation of the line that passes through the points $(-3, 5)$ and $(2, 15)$?

   **A**  $y = 2x + 11$    **B**  $y = 2x - 11$

   **C**  $y = -2x + 11$    **D**  $y = \frac{1}{2}x + 11$

   **E**  $y = -\frac{1}{2}x - 11$

3. *Multiple Choice*   Find an equation of the line through $(-3, 10)$ and $(2, -5)$.

   **A**  $y - 2 = 3(x + 5)$

   **B**  $y + 10 = 3(x + 3)$

   **C**  $y - 10 = -3(x - 3)$

   **D**  $y + 5 = -3(x - 2)$

   **E**  $y + 5 = 3(x - 2)$

4. *Multiple Choice*   Write an equation of the line shown in the graph.

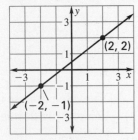

   **A**  $y = -\frac{3}{4}x + \frac{1}{2}$    **B**  $y = \frac{4}{3}x + \frac{1}{2}$

   **C**  $y = \frac{3}{4}x - \frac{1}{2}$    **D**  $y = \frac{4}{3}x - \frac{1}{2}$

   **E**  $y = \frac{3}{4}x + \frac{1}{2}$

5. *Multiple Choice*   Which lines are parallel?

   Line $d$ passes through $(2, 4)$ and $(-1, 6)$.

   Line $e$ passes through $(-3, -2)$ and $(5, 8)$.

   Line $f$ passes through $(6, -4)$ and $(3, -2)$.

   **A**  Lines $d$ and $e$

   **B**  Lines $d$ and $f$

   **C**  Lines $e$ and $f$

   **D**  All three are parallel

   **E**  None of these

6. *Multiple Choice*   What is the slope of the line through $(-7, 2)$ and $(-1, 2)$?

   **A**  $-\frac{1}{2}$    **B**  2    **C**  $\frac{-6}{0}$

   **D**  undefined    **E**  0

*Quantitative Comparison*   In Exercises 7–9, use the diagram and choose the statement that is true about the given numbers.

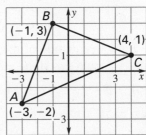

   **A**  The number in column A is greater.

   **B**  The number in column B is greater.

   **C**  The two numbers are equal.

   **D**  The relationship cannot be determined from the information given.

|   | Column A | Column B |
|---|---|---|
| 7. | The slope of line containing $\overline{AC}$ | The slope of line containing $\overline{BC}$ |
| 8. | The slope of line containing $\overline{AB}$ | The slope of line containing $\overline{BC}$ |
| 9. | The *y*-intercept of line containing $\overline{AB}$ | The *y*-intercept of line containing $\overline{BC}$ |

NAME _____ DATE _____

# *Standardized Test Practice*

For use with pages 291–297

**TEST TAKING STRATEGY** **Think positively during a test. This will help keep up your confidence and enable you to focus on each question.**

**1. *Multiple Choice*** Write $y = \frac{3}{7}x - 2$ in standard form with integer coefficients.

  **(A)** $7y = 3x - 14$   **(B)** $-3x + 7y = -2$

  **(C)** $-3x + 7y = 14$   **(D)** $3x - 7y = 14$

  **(E)** $3x - 7y = -14$

**2. *Multiple Choice*** Write in standard form an equation of the line that passes through the point $(-1, 5)$ and has a slope of $\frac{1}{2}$.

  **(A)** $x - 2y = -9$   **(B)** $x + 2y = 9$

  **(C)** $x + 2y = -11$   **(D)** $x + 2y = 11$

  **(E)** $x - 2y = -11$

**3. *Multiple Choice*** Write in standard form an equation of the line that passes through the point $(-8, -2)$ and has a slope of $-5$.

  **(A)** $5x + y = 42$   **(B)** $5x + y = -42$

  **(C)** $5x + y = -38$   **(D)** $5x - y = 38$

  **(E)** $5x + y = -18$

**4. *Multiple Choice*** Write in standard form an equation of the line that passes through the points $(-1, 4)$ and $(-7, -5)$.

  **(A)** $y - 4 = \frac{3}{2}(x + 1)$

  **(B)** $3x + 2y = 11$

  **(C)** $-3x + 2y = 11$

  **(D)** $-3x + 2y = -5$

  **(E)** $3x + 2y = -5$

**5. *Multiple Choice*** Write in standard form an equation of the line that passes through the points $(3, 0)$ and $(6, -8)$.

  **(A)** $8x + 3y = 24$   **(B)** $y = -\frac{8}{3}x + 8$

  **(C)** $8x + 3y = 8$   **(D)** $3x + 8y = 64$

  **(E)** $8x + 3y = -24$

**6. *Multiple Choice*** Write in standard form an equation of the horizontal line that passes through the point $(2, -8)$.

  **(A)** $y = 2$   **(B)** $x = 2$

  **(C)** $y = -8$   **(D)** $x = -8$

  **(E)** $y = 0$

**7. *Multiple Choice*** Write in standard form an equation of the vertical line in the graph.

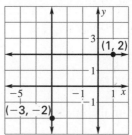

  **(A)** $y = -2$   **(B)** $x = -3$

  **(C)** $y = 2$   **(D)** $x = 1$

  **(E)** $y = -3$

***Quantitative Comparison*** In Exercises 8–10, choose the statement below that is true about the given numbers.

  **(A)** The number in column A is greater.

  **(B)** The number in column B is greater.

  **(C)** The two numbers are equal.

  **(D)** The relationship cannot be determined from the given information.

| | Column A | Column B |
|---|---|---|
| **8.** | The constant term in $3x + 2y = 7$ | $A$ in the standard form of $5x + 8y = 7$ |
| **9.** | The $x$-intercept of $3x - 2y = -10$ | The $y$-intercept of $2x + 5y = 6$ |
| **10.** | The slope of $y = -6$ | The slope of $12x + 3y = 16$ |

NAME _____ DATE _____

## *Standardized Test Practice*

**For use with pages 298–304**

**TEST TAKING STRATEGY** **Before you give up on a question, try to eliminate some of your choices so you can make an educated guess.**

*Multiple Choice* For Exercises 1–3, use the following information. The Changs currently have $14,300 in savings and add $2500 to it every year.

1. If $t = 0$ represents the current year, which linear model gives the Changs' savings in terms of $t$?

   Ⓐ  $S = 14{,}300t + 2500$

   Ⓑ  $S = 2500t + 14{,}300$

   Ⓒ  $S = 14{,}300t - 2500$

   Ⓓ  $S = 2500t - 14{,}300$

   Ⓔ  $S = 2500(t + 2500)$

2. What does the slope in the linear model represent?

   Ⓐ  The total savings amount

   Ⓑ  The number of years of saving

   Ⓒ  The current year

   Ⓓ  The current amount of savings

   Ⓔ  The annual increase in savings

3. Use the linear model from Exercise 1 to predict the Changs' savings 5 years from now.

   Ⓐ  $14,300     Ⓑ  $24,300

   Ⓒ  $26,800     Ⓓ  $74,000

   Ⓔ  $29,300

*Multiple Choice* For Exercises 4 and 5, use the following information. You have $105 in five and ten dollar bills.

4. Which equation relates the number of five dollar bills, $x$, and the number of ten dollar bills?

   Ⓐ  $x + y = 105$

   Ⓑ  $10x + 5y = 105$

   Ⓒ  $5x + 10y = 105$

   Ⓓ  $x - y = 105$

   Ⓔ  $105x + y = 1$

5. If you have 9 five dollar bills, how many ten dollar bills do you have?

   Ⓐ  6          Ⓑ  9          Ⓒ  3

   Ⓓ  12         Ⓔ  not enough information

6. *Multiple Choice* If $t = 0$ represents the year 1989, $t = $ __?__ represents the year 2011.

   Ⓐ  22         Ⓑ  20         Ⓒ  19

   Ⓓ  21         Ⓔ  23

7. *Multi-Step Problem* At 10 A.M. you are 8 miles from home. You jog at a rate of 6 miles per hour.

   a. Which linear model represents your distance from home $t$ hours after 10 A.M.?

   Ⓐ  $d = 8 + 6t$

   Ⓑ  $d = 8 - 6t$

   Ⓒ  $d = 6 + 8t$

   Ⓓ  $d = 6 - 8t$

   b. Use the linear model to complete the table.

   | Time (h), t       | 0 | 0.5 | 1 |
   |-------------------|---|-----|---|
   | Distance (mi), d  | ? | ?   | ? |

   c. At what time will you arrive at home?

**Algebra 1, Concepts and Skills**
Standardized Test Practice Workbook

# *Standardized Test Practice*

For use with pages 306–312

**TEST TAKING STRATEGY**   **Go back and check as much of your work as you can.**

1. *Multiple Choice*   What is an equation of the line that passes through points $(-2, 5)$ and $(3, 1)$?

   Ⓐ  $y = \frac{4}{5}x + \frac{17}{5}$   Ⓑ  $y = -\frac{4}{5}x + \frac{17}{5}$

   Ⓒ  $y = \frac{4}{5}x - \frac{17}{5}$   Ⓓ  $y = \frac{1}{2}x - \frac{3}{2}$

   Ⓔ  $y = \frac{1}{2}x + \frac{3}{2}$

2. *Multiple Choice*   What is an equation of the line that passes through point $(5, 2)$ and is perpendicular to $y = -2x + 1$?

   Ⓐ  $y = \frac{1}{2}x + \frac{3}{2}$   Ⓑ  $y = -2x + 12$

   Ⓒ  $y = -\frac{1}{2}x + \frac{9}{2}$   Ⓓ  $y = \frac{1}{2}x - \frac{1}{2}$

   Ⓔ  $y = \frac{1}{2}x + \frac{1}{2}$

3. *Multiple Choice*   What is the slope of a line perpendicular to the line $y = \frac{1}{2}x - 6$?

   Ⓐ  2   Ⓑ  $-\frac{1}{2}$   Ⓒ  $-2$

   Ⓓ  $\frac{1}{2}$   Ⓔ  $\frac{1}{6}$

4. *Multiple Choice*   Write an equation of the line shown in the graph.

   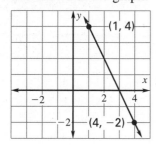

   Ⓐ  $y = -\frac{1}{2}x + 8$   Ⓑ  $y = \frac{1}{2}x - 6$

   Ⓒ  $y = \frac{1}{2}x + 6$   Ⓓ  $y = 2x + 6$

   Ⓔ  $y = -2x + 6$

5. *Multiple Choice*   Which pair of lines are perpendicular?

   Ⓐ  $y = 2x + 3; y = -2x + 6$

   Ⓑ  $y = \frac{1}{3}x; y = -3x + 2$

   Ⓒ  $y = \frac{2}{3}x + \frac{1}{2}; y = \frac{2}{3}x - 2$

   Ⓓ  $y = -6x + 1; y = -\frac{1}{6}x - 1$

   Ⓔ  $y = 3; y = x + 1$

6. *Multiple Choice*   Which lines are perpendicular?

   Line $d$ passes through $(2, 4)$ and $(-1, 6)$.

   Line $e$ passes through $(-3, -2)$ and $(5, 8)$.

   Line $f$ passes through $(2, 10)$ and $(7, 6)$.

   Ⓐ  Lines $d$ and $e$   Ⓑ  Lines $d$ and $f$

   Ⓒ  Lines $e$ and $f$

   Ⓓ  All three are perpendicular.

   Ⓔ  None of these

7. *Multiple Choice*   Which is an equation of a line perpendicular to the line $y = -\frac{3}{2}x + \frac{1}{2}$?

   Ⓐ  $y = -\frac{2}{3}x + 6$   Ⓑ  $y = \frac{2}{3}x + 3$

   Ⓒ  $y = \frac{3}{2}x - \frac{1}{2}$   Ⓓ  $y = -\frac{2}{3}x - \frac{1}{2}$

   Ⓔ  $y = -\frac{3}{2}x - 2$

8. *Multi-Step Problem*   The triangle shown is a right triangle.

   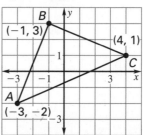

   a. Find the equations of the lines that form side $\overline{AB}$, side $\overline{BC}$, and side $\overline{AC}$.

   b. Which two sides of the triangle form the right angle? How can you prove it?

**Algebra 1, Concepts and Skills**
Standardized Test Practice Workbook

NAME _____ DATE _____

# Standardized Test Practice

For use with pages 323–328

**TEST TAKING STRATEGY** **Learn as much as you can about a test ahead of time, such as the types of questions and the topics that the test will cover.**

1. **Multiple Choice** Describe the inequality $y > 8$.

Ⓐ All real numbers greater than 8

Ⓑ All real numbers greater than or equal to 8

Ⓒ All real numbers greater than or equal to 9

Ⓓ All real numbers less than 8

Ⓔ All real numbers less than or equal to 8

2. **Multiple Choice** Describe the solution of $x - 8 > 3$.

Ⓐ All real numbers greater than $-5$

Ⓑ All real numbers less than $-5$

Ⓒ All real numbers greater than 11

Ⓓ All real numbers less than 11

Ⓔ None of these

3. **Multiple Choice** Choose the solution of $-5 + x > 10$.

Ⓐ $x > 5$  Ⓑ $x > -5$

Ⓒ $x > 15$  Ⓓ $x < 15$

Ⓔ $x < 5$

4. **Multiple Choice** Choose the solution of $2 + x < -5$.

Ⓐ $x \le -7$  Ⓑ $x < 7$

Ⓒ $x < -7$  Ⓓ $-7 \le x$

Ⓔ $7 \le x$

5. **Multiple Choice** The coldest temperature recorded in your city was $-6°C$. Choose the inequality describing all of the recorded temperatures in your city.

Ⓐ $t < -6$  Ⓑ $t > -6$

Ⓒ $t \le -6$  Ⓓ $t \ge -6$

Ⓔ None of these

6. **Multiple Choice** Which graph represents the solution of $x + 7 > 3$?

Ⓐ

Ⓑ

Ⓒ

Ⓓ

Ⓔ

7. **Multiple Choice** Which graph represents the solution of $1 \le -1 + x$?

Ⓐ

Ⓑ

Ⓒ

Ⓓ

Ⓔ

8. **Multi-Step Problem** At an auction, the minimum bid for antique chair is $250. Let $b$ represent any bid for the chair.

a. Write an inequality that describes $b$.

b. Is 180 a solution of the inequality?

c. Name three solutions of the inequality.

**Algebra 1, Concepts and Skills**
Standardized Test Practice Workbook

NAME _____ DATE _____

# *Standardized Test Practice*

For use with pages 330–335

**TEST TAKING STRATEGY   Learn as much as you can about a test ahead of time, such as the types of questions and the topics that the test will cover.**

**1.** *Multiple Choice*   Describe the solution of $2x \leq 14$.

Ⓐ  All real numbers less than 7

Ⓑ  All real numbers less than or equal to 7

Ⓒ  All real numbers greater than or equal to 7

Ⓓ  All real numbers less than 12

Ⓔ  None of these

**2.** *Multiple Choice*   Describe the solution of $3x \leq -9$.

Ⓐ  All real numbers greater than −3

Ⓑ  All real numbers greater than or equal to −3

Ⓒ  All real numbers greater than or equal to −6

Ⓓ  All real numbers less than −3

Ⓔ  All real numbers less than or equal to −3

**3.** *Multiple Choice*   Choose the solution of $-x \leq 1$.

Ⓐ  $x < 2$     Ⓑ  $x < -1$

Ⓒ  $x < 1$     Ⓓ  $x \geq -1$

Ⓔ  $x \leq 1$

**4.** *Multiple Choice*   Choose the solution of $-\frac{x}{7} \geq 3$.

Ⓐ  $x \geq -4$     Ⓑ  $x \geq -21$

Ⓒ  $x \leq -21$     Ⓓ  $x \leq -\frac{3}{7}$

Ⓔ  $x \leq -4$

**5.** *Multiple Choice*   Which graph represents the solution of $\frac{x}{6} > \frac{1}{2}$?

Ⓐ

Ⓑ

Ⓒ

Ⓓ

Ⓔ

**6.** *Multiple Choice*   Which graph represents the solution of $-3x \leq 6$?

Ⓐ

Ⓑ

Ⓒ

Ⓓ

Ⓔ

*Quantitative Comparison*   In Exercises 7–9, choose the statement that is true about the given numbers.

Ⓐ  The number in column A is greater.

Ⓑ  The number in column B is greater.

Ⓒ  The two numbers are equal.

Ⓓ  The relationship cannot be determined from the given information.

| | Column A | Column B |
|---|---|---|
| **7.** | $x < 6$ | $x \geq 6$ |
| **8.** | $x - 10 \geq -3$ | $2x < 10$ |
| **9.** | $-\frac{x}{3} \leq 5$ | $x \leq 0$ |

# Standardized Test Practice

For use with pages 336–341

**TEST TAKING STRATEGY**   Be aware of how much time you have left, but keep focused on your work.

1. *Multiple Choice*   Which inequality is equivalent to $3x - 2 \leq 7$?

   Ⓐ $x \geq 3$      Ⓑ $x \leq 3$

   Ⓒ $x \leq \frac{5}{3}$      Ⓓ $x \geq \frac{5}{3}$

   Ⓔ $x \leq 6$

2. *Multiple Choice*   Which inequality is equivalent to $5 - 8x \geq -11$?

   Ⓐ $x \leq 2$      Ⓑ $x \geq 2$

   Ⓒ $x \leq -2$      Ⓓ $x \leq \frac{3}{4}$

   Ⓔ $x \geq \frac{3}{4}$

3. *Multiple Choice*   Describe the solution of the inequality $-\frac{1}{2}x - 4 \leq 6$.

   Ⓐ All real numbers greater than or equal to $-20$

   Ⓑ All real numbers greater than or equal to $20$

   Ⓒ All real numbers less than or equal to $-20$

   Ⓓ All real numbers less than or equal to $-5$

   Ⓔ All real numbers greater than or equal to $-5$

4. *Multiple Choice*   Which inequality is equivalent to $x - 5 \leq 3x + 7$?

   Ⓐ $x \geq 6$      Ⓑ $x \geq -1$

   Ⓒ $x \leq -6$      Ⓓ $x \geq -6$

   Ⓔ $x \leq -1$

5. *Multiple Choice*   Which inequality is equivalent to $6 - x > 4x + 9$?

   Ⓐ $x < 3$      Ⓑ $x > 3$

   Ⓒ $x < -\frac{3}{5}$      Ⓓ $x > -\frac{3}{5}$

   Ⓔ $x < -3$

6. *Multiple Choice*   It costs $20 to golf at a local course. A season pass costs $250. Which inequality represents the number of games you need to play to justify buying a season pass?

   Ⓐ $x \geq 12$      Ⓑ $x < 13$

   Ⓒ $x \geq 13$      Ⓓ $x \leq 12$

   Ⓔ None of these

7. *Multi-Step Problem*   You are working on your school's yearbook. It takes at least 3 hours to design and lay out a page.

   a. Write an inequality that describes the number of hours it would take to lay out a 220 page yearbook.

   b. You work 4 hours per week. Write and solve an inequality describing the number of weeks it would take you to complete the yearbook if you were working alone.

   c. Write and solve an inequality to describe the number of people that would be necessary to complete the yearbook in 20 weeks or less if each person averages 5 hours of work per week.

**Algebra 1, Concepts and Skills**
Standardized Test Practice Workbook

NAME _____ DATE _____

# *Standardized Test Practice*

**For use with pages 342–347**

**TEST TAKING STRATEGY**   **Some questions involve more than one step. Reading too quickly might lead to mistaking the answer to a preliminary step for your final answer.**

1. *Multiple Choice*   Which inequality represents all real numbers greater than 4 and less than or equal to 7?

   (A)  $4 < x \le 7$  (B)  $4 \le x \le 7$

   (C)  $4 < x < 7$  (D)  $4 > x$ or $x \le 7$

   (E)  $4 < x$ or $x \ge 7$

2. *Multiple Choice*   Describe the solution of the inequality $-2 < 3x + 1 < 10$.

   (A)  All real numbers less than 3 and greater than $-2$

   (B)  All real numbers less than 10 or greater than $-1$

   (C)  All real numbers less than 3 and greater than $-1$

   (D)  All real numbers less than 9 or greater than $-1$

   (E)  None of these

3. *Multiple Choice*   Describe the solution of the inequality $8 \le x - 3 < 12$.

   (A)  All real numbers less than 9 and greater than or equal to 5

   (B)  All real numbers less than 15 or greater than or equal to 11

   (C)  All real numbers less than 15 and greater than or equal to 11

   (D)  All real numbers less than 9 or greater than or equal to 5

   (E)  None of these

4. *Multiple Choice*   Choose the solution of the inequality $6 \le -2x < 14$.

   (A)  $-7 < x \le -3$  (B)  $-3 \le x < -7$

   (C)  $-3 \ge x > -7$  (D)  A and C

   (E)  A and B

5. *Multiple Choice*   Which graph represents the solution of $-3 \le 4x + 1 < 9$?

   (A)

   (B)

   (C)

   (D)

   (E)

6. *Multiple Choice*   Which graph represents the solution of $3 < -2x + 1 < 7$?

   (A)

   (B)

   (C)

   (D)

   (E)  None of these

*Quantitative Comparison*   In Exercises 7–9, choose the statement below that is true about the given numbers. In each exercise, x represents an integer.

   (A)  The number in column A is greater.

   (B)  The number in column B is greater.

   (C)  The two numbers are equal.

   (D)  The relationship cannot be determined from the information given.

|   | Column A | Column B |
|---|---|---|
| 7. | $-4 < 2x < 0$ | $x > -1$ and $x < 1$ |
| 8. | $6 < 2x + 10 < 14$ | $x > 2$ and $x \le 6$ |
| 9. | $-13 \le 3x + 5 < -1$ | $x < 0$ and $x > -10$ |

Chapter 6

NAME _____    DATE _____

# *Standardized Test Practice*

For use with pages 348–353

**TEST TAKING STRATEGY**   **Some questions involve more than one step. Reading too quickly might lead to mistaking the answer to a preliminary step for your final answer.**

1. *Multiple Choice*   Which inequality represents all real numbers greater than or equal to 5 or less than $-2$?

Ⓐ  $5 \le x < -2$  Ⓑ  $x \ge 5$ or $x \le -2$

Ⓒ  $5 \ge x > -2$  Ⓓ  $x \le 5$ or $x < -2$

Ⓔ  $5 > x > -2$

2. *Multiple Choice*   Describe the solution of the compound inequality $-2x + 5 < 9$ or $-5x + 2 > 17$.

Ⓐ  All real numbers less than $-3$ and greater than 2

Ⓑ  All real numbers less than 3 and greater than $-2$

Ⓒ  All real numbers less than 3 and greater than 2

Ⓓ  All real numbers less than $-3$ or greater than $-2$

Ⓔ  All real numbers less than $-\frac{19}{5}$ and greater than $-7$

3. *Multiple Choice*   Choose the solution of the compound inequality $2x - 5 > 3$ or $2 - 3x > 5$.

Ⓐ  $x < -1$ or $x > 4$
Ⓑ  $x < 5$ or $x > 3$
Ⓒ  $x < 4$ or $x > -1$
Ⓓ  $x < 3$ or $x > 5$
Ⓔ   $-1 < x < 4$

4. *Multiple Choice*   Choose the solution of the compound inequality $5x \ge 20$ or $5 + 4x < -1.9$

Ⓐ  $x \ge -4$ or $x < -6$

Ⓑ  $x \ge -4$ or $x < 6$

Ⓒ  $x \ge -4$ and $x < 6$

Ⓓ  $x \ge 4$ or $x < -6$

Ⓔ  None of these

5. *Multiple Choice*   Which graph represents the solution of the compound inequality $3 - x < 1$ or $4x + 3 \le -1$?

Ⓐ

Ⓑ

Ⓒ

Ⓓ

Ⓔ  None of these

6. *Multiple Choice*   Which graph represents the solution of the inequality $4x + 1 < 13$ or $3x - 8 \ge 10$?

Ⓐ

Ⓑ

Ⓒ

Ⓓ

Ⓔ  None of these

7. *Multi-Step Problem*   A ball is thrown straight up. Its initial velocity is 48 feet per second. Its velocity $v$ in feet per second is given by $v = -32t + 48$.

a. Complete the table.

| t (sec) | 0 | 0.5 | 1 | 1.5 | 2 | 2.5 |
|---------|---|-----|---|-----|---|-----|
| v (ft/sec) | ? | ? | ? | ? | ? | ? |

b. Describe the results.

c. Write a compound inequality to show the values of $t$ for which the velocity is greater than 32 or less than $-32$.

NAME _____ DATE _____

# *Standardized Test Practice*

**For use with pages 355–360**

**TEST TAKING STRATEGY   Work as fast as you can through the easier problems, but not so fast that you are careless.**

1. *Multiple Choice*   Which numbers are solutions to the absolute-value equation $|x + 2| - 3 = 8$?

    Ⓐ   6 and $-10$   Ⓑ   9 and $-13$

    Ⓒ   9 and $-9$   Ⓓ   $-9$ and $-13$

    Ⓔ   3 and $-7$

2. *Multiple Choice*   Which numbers are solutions to the absolute-value equation $12 + |3x - 1| = 19$?

    Ⓐ   2 and $-2$   Ⓑ   $\frac{8}{3}$ and $-\frac{8}{3}$

    Ⓒ   $\frac{32}{3}$ and $-10$   Ⓓ   $\frac{8}{3}$ and $-2$

    Ⓔ   10 and $-10$

3. *Multiple Choice*   Which numbers are solutions to the absolute-value equation $|1 - x| = 2$?

    Ⓐ   $-1$ and 2   Ⓑ   $-2$ and 2

    Ⓒ   $-1$ and $-2$   Ⓓ   $-1$ and 3

    Ⓔ   None of these

4. *Multiple Choice*   Which numbers are solutions to the absolute-value equation $7 + |x - 6| = 3$?

    Ⓐ   $-2$ and $-10$   Ⓑ   $-2$ and 10

    Ⓒ   2 and 10   Ⓓ   2 and $-10$

    Ⓔ   None of these

5. *Multiple Choice*   Which numbers are solutions to the absolute-value equation $\left|\dfrac{x}{4}\right| = 0$?

    Ⓐ   0   Ⓑ   4

    Ⓒ   0 and 4   Ⓓ   4 and $-4$

    Ⓔ   None of these

6. *Multiple Choice*   Which graph represents the solution of $|2x - 5| = 3$?

    Ⓐ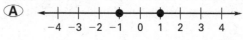

    Ⓑ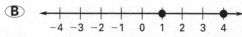

    Ⓒ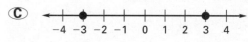

    Ⓓ

    Ⓔ   None of these

*Quantitative Comparison*   In Exercises 7–9, choose the statement below that is true about the given numbers.

Ⓐ   The number in column A is greater.

Ⓑ   The number in column B is greater.

Ⓒ   The two numbers are equal.

Ⓓ   The relationship cannot be determined from the given information.

|     | Column A | Column B |
| --- | --- | --- |
| 7. | $|x| = 6$ | $|x| = 0$ |
| 8. | $|x - 1| = 1$ | $|x - 4| = 1$ |
| 9. | $|2x - 3| = 1$ | $|2x + 3| = 1$ |

**Algebra 1, Concepts and Skills**
Standardized Test Practice Workbook

# *Standardized Test Practice*

**For use with pages 361–366**

**TEST TAKING STRATEGY   Work as fast as you can through the easier problems, but not so fast that you are careless.**

1. *Multiple Choice*   Which number is a solution to the absolute-value inequality $|6x - 7| < 2$?

   Ⓐ $\frac{2}{3}$         Ⓑ  0.5

   Ⓒ  $-1$         Ⓓ  1.23

   Ⓔ   None of these

2. *Multiple Choice*   Which number is a solution to the absolute-value inequality $|x + 1| < 1$?

   Ⓐ  $-1$         Ⓑ  $-3$

   Ⓒ  0         Ⓓ  1

   Ⓔ   None of these

3. *Multiple Choice*   Which graph represents the solution of $|3x - 6| \geq 9$?

4. *Multiple Choice*   Which graph represents the solution of $|2x - 7| + 3 < 12$?

5. *Multiple Choice*   Which graph represents the solution of $|5x - 2| - 8 \geq 9$?

6. *Multiple Choice*   Which of the following is a solution to the absolute-value inequality $|x - 6| \leq 3$?

   Ⓐ  10         Ⓑ  1         Ⓒ  3

   Ⓓ  12         Ⓔ  2

*Quantitative Comparison*   In Exercises 7–9, choose the statement below that is true about the given numbers.

   Ⓐ   The number in column A is greater.

   Ⓑ   The number in column B is greater.

   Ⓒ   The two numbers are equal.

   Ⓓ   The relationship cannot be determined from the given information.

| | Column A | Column B |
|---|---|---|
| 7. | $|x + 2| < 1$ | $|x| < 1$ |
| 8. | $|x - 7| < 2$ | $|2x + 3| < 7$ |
| 9. | $|4x - 2| < 6$ | $|2x - 1| - 1 < 2$ |

**Algebra 1, Concepts and Skills**
Standardized Test Practice Workbook

NAME _____ DATE _____

# *Standardized Test Practice*

**For use with pages 367–374**

**TEST TAKING STRATEGY**  **If you can, check your answer using a different method than you used originally to avoid making the same mistake twice.**

**1.** *Multiple Choice*  Which point is a solution of $3x - 5y \geq 8$?

ⓐ  (1, 1)       Ⓑ  (4, 1)

Ⓒ  (−2, −2)  Ⓓ  (2, −1)

Ⓔ  (−1, 1)

**2.** *Multiple Choice*  Which point is a solution of $4x - 8y < 14$?

ⓐ  (0, 2)       Ⓑ  (3, −1)

Ⓒ  (0, −2)     Ⓓ  (5, −2)

Ⓔ  (−3, −4)

**3.** *Multiple Choice*  Which point is not a solution of $11x - 6 < y$?

ⓐ  (−1, 3)     Ⓑ  (2, 10)

Ⓒ  (0, 3)       Ⓓ  (−4, −1)

Ⓔ  (1, 7)

**4.** *Multiple Choice*  Choose the inequality whose solution is shown in the graph.

ⓐ  $x > -1$

Ⓑ  $y > -1$

Ⓒ  $x \geq -1$

Ⓓ  $y \geq -1$

Ⓔ  $x - 1 \geq -1$

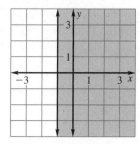

**5.** *Multiple Choice*  Choose the inequality whose solution is shown in the graph.

ⓐ  $y < -3x + 2$

Ⓑ  $y > -3x + 2$

Ⓒ  $y > 3x + 2$

Ⓓ  $y \geq -3x + 2$

Ⓔ  $y \leq -3x + 2$

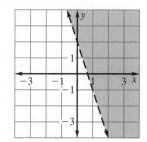

**6.** *Multiple Choice*  Choose the inequality that the graph represents.

ⓐ  $y > \frac{2}{3}x - 1$

Ⓑ  $y < \frac{2}{3}x - 1$

Ⓒ  $y \leq -\frac{2}{3}x - 1$

Ⓓ  $y \leq -\frac{2}{3}x + 1$

Ⓔ  $y \geq \frac{2}{3}x - 1$

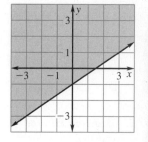

**7.** *Multi-Step Problem*  You are in charge of a dinner-dance for school. You have a budget of $500 for dinners. There are 3 choices of meals: a fish dinner for $8.50, a steak dinner for $10.50, and a vegetarian dinner for $7.50. Two people order the vegetarian meal.

**a.** Write an inequality to model the different combinations of fish and steak dinners that might be purchased.

**b.** Graph the inequality.

**c.** If only 20 fish dinners are available, how many steak dinners can be purchased?

**d.** Does every point on the graph represent a reasonable real-life solution? Explain.

**Algebra 1, Concepts and Skills**
Standardized Test Practice Workbook

# *Cumulative Standardized Test Practice*

**For use after Chapters 1–6**

**1.** *Multiple Choice*   Evaluate $\frac{2}{3}x$ when $x = \frac{9}{4}$.

  **A** $\frac{11}{7}$      **B** $\frac{3}{2}$      **C** $\frac{2}{3}$

  **D** $\frac{18}{7}$      **E** $\frac{4}{9}$

**2.** *Multiple Choice*   Evaluate the expression $(x - y)^3$ when $x = 3$ and $y = -2$.

  **A** 1      **B** $-1$      **C** 125

  **D** 25      **E** $-125$

**3.** *Multiple Choice*   Evaluate $(-2x)^3$ when $x = 3$.

  **A** $-216$      **B** $-54$      **C** 125

  **D** $-125$      **E** 216

**4.** *Multiple Choice*   What is the value of $\dfrac{3 \cdot 4 + 6^2}{16 - 3^2 \cdot 2}$?

  **A** 24      **B** $-12$      **C** $\frac{13}{4}$

  **D** $\frac{48}{5}$      **E** $-24$

*Quantitative Comparison*   In Exercises 5–7, choose the statement below that is true about the given number.

  **A**   The number in column A is greater.

  **B**   The number in column B is greater.

  **C**   The two numbers are equal.

  **D**   The relationship cannot be determined from the information given.

|   | Column A | Column B |
|---|----------|----------|
| **5.** | The value of $x$ when $8x - 3 = 29$ | The value of $x$ when $18 = \frac{1}{2}x + 12$ |
| **6.** | Nine times a number decreased by 8 is 19. | Eight times a number equals three squared plus seven. |
| **7.** | $3x^2 - 8$ when $x = -2$ | $118 - 4x^2$ when $x = 9$ |

**8.** *Multiple Choice*   Evaluate the expression $-|-6| + 3$.

  **A** 3      **B** 9      **C** 0

  **D** $-3$      **E** $-9$

**9.** *Multiple Choice*   Evaluate the expression $7 + x + (-2)$ when $x = -8$.

  **A** 13      **B** 17      **C** $-3$

  **D** $-13$      **E** 3

**10.** *Multiple Choice*   Evaluate the expression $9x - (2x)^3 + x$ when $x = -2$.

  **A** $-44$      **B** $-48$      **C** $-78$

  **D** 80      **E** 44

**11.** *Multiple Choice*   Simplify the expresion $8x^2 - x^2 + 7x - 2x$.

  **A** $9x^2 - 9x$   **B** $7x^2 - 2x - 7$

  **C** $9x^2 + 5x$   **D** $7x^2 - 5x$

  **E** $7x^2 + 5x$

**12.** *Multiple Choice*   Find the missing lengths if the perimeter is 21.

  **A** 2, 4

  **B** 3, 6

  **C** 5, 10

  **D** 6, 12      **E** 4, 8

Chapter 6

**Quantitative Comparison**   In Exercises 13–15, choose the statement below that is true about the given numbers.

**(A)**   The solution of column A is greater.

**(B)**   The solution of column B is greater.

**(C)**   The two solutions are equal.

**(D)**   The relationship cannot be determined from the information given.

| | Column A | Column B |
|---|---|---|
| **13.** | $2x + |-3| = 17$ | $-|-7| + 3x = 19$ |
| **14.** | $-12 = \frac{x}{4}$ | $-\frac{2}{3}x = 32$ |
| **15.** | $4x - 2(x - 3) = 17$ | $\frac{1}{2}(x - 4) = 12$ |

**16. Multiple Choice**   Solve $5x - 6 = 4x + 3$.

**(A)** 9   **(B)** 8   **(C)** 7

**(D)** 6   **(E)** 5

**17. Multiple Choice**   Solve the equation $\frac{1}{3}(9x - 15) - 2 = 2x + 5$.

**(A)** 12   **(B)** 8   **(C)** 22

**(D)** $-22$   **(E)** $-8$

**18. Multiple Choice**   Solve the equation $\frac{1}{4}(4x - 4) = \frac{1}{5}(5x - 10)$.

**(A)** 0   **(B)** 2   **(C)** $-3$

**(D)** 1   **(E)** No solution

**19. Multi-Step Problem**   Your cable TV company charges $32 a month for basic cable and one premium channel. Additional premium channels cost $12 per month.

**a.** Let $x$ represent the number of premium channels you subscribe to. Write an expression that models the total cost of your cable bill.

**b.** Your bill came to $56 last month. How many premium channels did you order?

**20. Multiple Choice**   Solve the equation $2.3(-1.2x + 8.2) = 2.1x$. Round to the nearest tenth.

**(A)** $-28.6$   **(B)** 28.6   **(C)** 3.8

**(D)** 3.9   **(E)** $-3.9$

**21. Multiple Choice**   A triangle has an area of 30 square inches and a base of 10 inches. Find its height. (Use the formula $A = \frac{1}{2}bh$.)

**(A)** 1.5 in.   **(B)** 3 in.

**(C)** 9 in.   **(D)** 20 in.

**(E)** 6 in.

**22. Multiple Choice**   Rewrite the equation $5x + 3y = 12$ so that $y$ is a function of $x$.

**(A)** $y = -\frac{5}{3}x + 4$   **(B)** $y = \frac{5}{3}x + 4$

**(C)** $y = \frac{3}{5}x + 4$   **(D)** $y = 2x + 9$

**(E)** $y = -2x + 9$

**23. Multiple Choice**   In which quadrant of the coordinate plane is point $(-3, 6)$?

**(A)** I   **(B)** II   **(C)** III

**(D)** IV   **(E)** II or IV

**24. Multiple Choice**   Choose the ordered pair that is a solution of $3x - 2y = 8$.

**(A)** $(-4, -2)$   **(B)** $(2, 1)$   **(C)** $(8, 6)$

**(D)** $(2, -1)$   **(E)** $(-2, -8)$

**25. Multiple Choice**   What is an equation of the line shown in the graph?

**(A)** $-5x + 3y = 15$

**(B)** $-5x + 3y = -15$

**(C)** $5x - 3y = 15$

**(D)** $5x + 3y = 15$

**(E)** $5x - 3y = -15$

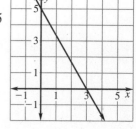

NAME _____ DATE _____

# Cumulative Standardized Test Practice

**For use after Chapters 1–6**

26. **Multiple Choice**   What is the $y$-intercept of the graph shown in Exercise 25?

    Ⓐ 3      Ⓑ 5      Ⓒ $-5$

    Ⓓ $-3$      Ⓔ 4

27. **Multiple Choice**   Find the slope of the line passing through the points $(7, 2)$ and $(8, 6)$.

    Ⓐ $-4$      Ⓑ $\frac{1}{4}$      Ⓒ 4

    Ⓓ $-\frac{1}{4}$      Ⓔ $-6$

28. **Multiple Choice**   The variables $x$ and $y$ vary directly when $x = 10$, $y = 2$. Which equation correctly relates $x$ and $y$?

    Ⓐ $y = 10x$      Ⓑ $y = 5x$

    Ⓒ $y = \frac{1}{10}x$      Ⓓ $y = \frac{1}{5}x$

    Ⓔ $y = x - 8$

29. **Multiple Choice**   Choose the pair of lines which are parallel.

    Ⓐ $y = \frac{2}{3}x + 1$, $y = \frac{3}{2}x + 6$

    Ⓑ $y = -2x + 3$, $y = 2x + 3$

    Ⓒ $y = \frac{1}{2}x - 1$, $y = \frac{1}{3}(x - 2)$

    Ⓓ $y = \frac{1}{5}x + 2$, $y = -5x - 6$

    Ⓔ $y = \frac{2}{7}x$, $y = \frac{2}{7}x + 2$

**Quantitative Comparison**   In Exercises 30–33, choose the statement below that is true about the given quantities.

Ⓐ   The value of column A is greater.

Ⓑ   The value of column B is greater.

Ⓒ   The two values are equal.

Ⓓ   The relationship cannot be determined from the information given.

|  | Column A | Column B |
|---|---|---|
| 30. | The constant of variation in $y = -3x$ | The constant of variation in $y = \frac{1}{2}x$ |
| 31. | The slope of $y = 13x + 6$ | The slope of $y = 6x + 13$ |
| 32. | The $y$-intercept of $y = -3x + \frac{1}{2}$ | The $y$-intercept of $y = \frac{1}{3}x + \frac{1}{2}$ |
| 33. | $f(x) = -2x + \frac{1}{2}$ when $x = -3$ | $f(x) = \frac{2}{3}x - 1$ when $x = 6$ |

34. **Multiple Choice**   What is the equation of the line that passes through the point $(2, -4)$ and has a slope of $\frac{1}{2}$?

    Ⓐ $y = \frac{1}{2}x - 5$      Ⓑ $y = \frac{1}{2}x + 5$

    Ⓒ $y = \frac{1}{2}x - 6$      Ⓓ $y = \frac{1}{2}x - 3$

    Ⓔ $y = \frac{1}{2}x + 6$

35. **Multiple Choice**   What is the slope of the line perpendicular to the line $y = \frac{2}{5}x + 10$?

    Ⓐ $\frac{2}{5}$      Ⓑ $-\frac{2}{5}$      Ⓒ $\frac{5}{2}$

    Ⓓ $-\frac{5}{2}$      Ⓔ $-\frac{1}{10}$

36. **Multiple Choice**   Which equation in point-slope form passes through the point $(5, -2)$ and has a slope of $-2$?

    Ⓐ $y = -2x - 12$

    Ⓑ $y + 2 = -2(x - 5)$

    Ⓒ $y = -2x + 8$

    Ⓓ $y - 2 = -2(x - 5)$

    Ⓔ $y - 2 = -2(x + 5)$

**37.** *Multiple Choice* Find the standard form of the equation of the line that passes through the points $(3, 5)$ and $(-1, -2)$.

  **A** $7x - 4y = 4$     **B** $7x - 4y = 1$

  **C** $y = \frac{7}{4}x - \frac{1}{4}$     **D** $7x - 4y = -5$

  **E** $y = \frac{7}{4}x - 5$

**38.** *Multi-Step Problem* The table shows the average cost of a large, one item pizza.

| Year | 1990 | 1992 | 1994 | 1996 | 1998 |
|------|------|------|------|------|------|
| Cost | $5.60 | $6.20 | $6.80 | $7.50 | $8.30 |

  **a.** Draw a scatter plot of the price of a pizza in terms of the year $t$. Let $t$ be the number of years since 1990.

  **b.** What does the scatter plot indicate?

**39.** *Multiple Choice* Which graph represents the solution of $-2x + 3 \geq 7$?

  **A**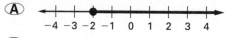

  **B**

  **C**

  **D**

  **E**

**40.** *Multiple Choice* Which inequality is equivalent to $x - 5 \leq 2x - 6$?

  **A** $x \leq 1$   **B** $x \geq -1$   **C** $x \geq 1$

  **D** $x \geq 11$   **E** $x \leq 11$

**41.** *Multiple Choice* Choose the solution of the inequality $6 < 2x - 3 \leq 11$.

  **A** $\frac{3}{2} > x \geq 7$   **B** $\frac{9}{2} < x \leq 7$

  **C** $\frac{9}{2} > x \geq 7$   **D** $\frac{3}{2} < x \leq \frac{9}{2}$

  **E** $\frac{3}{2} < x \leq 7$

**42.** *Multiple Choice* Which numbers are solutions of the absolute-value equation $|x - 1| - 2 = 6$?

  **A** 9 and $-7$     **B** $-9$ and 7

  **C** 5 and $-3$     **D** 7 and 8

  **E** $-5$ and $-3$

**43.** *Multiple Choice* Which point is not a solution of $y \geq 3x - 5$?

  **A** $(1, 6)$   **B** $(2, 1)$   **C** $(3, 2)$

  **D** $(-2, 3)$   **E** $(-1, 1)$

**44.** *Multi-Step Problem* Sean has reports to copy. He has already copied 28 pages. The copy machine takes 10 seconds to copy a page.

  **a.** How many pages can the copy machine copy in a minute?

  **b.** Write a linea model to relate the total number of pages copied, $y$, to the number of minutes $t$.

  **c.** Use the linear model to predict the number of pages copied in 12 minutes.

**Chapter 6**

NAME _____ DATE _____

# Standardized Test Practice

**For use with pages 389–395**

**TEST TAKING STRATEGY** **Some questions involve more than one step. Reading too quickly might lead to mistaking the answer to a preliminary step for your final answer.**

1. *Multiple Choice* Which point represents the solution of the system of linear equations?

$y = 2x - 3$

$y = -x - 2$

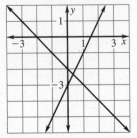

Ⓐ $(0, -3)$ Ⓑ $(0, -2)$ Ⓒ $\left(\frac{2}{3}, -\frac{5}{3}\right)$

Ⓓ $(1, 3)$ Ⓔ $\left(\frac{1}{3}, -\frac{7}{3}\right)$

2. *Multiple Choice* Which point represents the solution of the system of linear equations?

$y = -\frac{1}{2}x + 2$

$3x - 2y = 4$

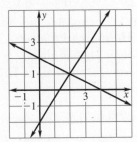

Ⓐ $(1, 2)$ Ⓑ $(2, 0)$ Ⓒ $(2, 1)$

Ⓓ $(0, -2)$ Ⓔ $(0, 2)$

3. *Multiple Choice* What is the solution of the system of linear equations?

$y = \frac{3}{2}x + 1$ $\quad y = \frac{3}{4}x - 2$

Ⓐ $(4, 5)$ Ⓑ $(-4, -5)$

Ⓒ $(-4, 5)$ Ⓓ $(4, -5)$

Ⓔ $(0, -2)$

4. *Multiple Choice* What is the solution of the system of linear equations?

$y = -\frac{2}{3}x + 1$ $\quad y = \frac{2}{3}x - 3$

Ⓐ $(3, 1)$ Ⓑ $(-3, -1)$

Ⓒ $(3, -1)$ Ⓓ $(-3, 1)$

Ⓔ $(3, 0)$

5. *Multiple Choice* If $y = 2x + 5$ and $4x - y = 3$, then $x = $ __?__ .

Ⓐ 1 Ⓑ 2 Ⓒ 3

Ⓓ 4 Ⓔ 5

6. *Multiple Choice* Riding the bus to and from work costs you $5.50 round trip. You are thinking about buying a used car for $2500. Gas and maintenance will cost you about $.50 per trip. How many trips to work will it take to break even?

Ⓐ 510 Ⓑ 417 Ⓒ 450

Ⓓ 500 Ⓔ 525

7. *Multiple Choice* If $y = \frac{1}{2}x - 2$ and $x + 2y = 0$, then $x + y = $ __?__ .

Ⓐ 2 Ⓑ 1 Ⓒ 6

Ⓓ 3 Ⓔ $-1$

*Quantitative Comparison* In Exercises 8–10, solve the linear system. Then choose the statement below that is true about the solution of the system.

Ⓐ The value of $x$ is greater than the value of $y$.

Ⓑ The value of $y$ is greater than the value of $x$.

Ⓒ The values of $x$ and $y$ are equal.

Ⓓ The relationship cannot be determined from the given information.

8. $y = \frac{1}{2}x - 4$
   $y = -\frac{3}{2}x + 1$

9. $3x + 2y = 14$
   $y = -\frac{3}{2}x + 1$

10. $y = \frac{1}{2}x - 4$
    $x = -1$

**Algebra 1, Concepts and Skills**
Standardized Test Practice Workbook

NAME _____ DATE _____

# Standardized Test Practice

**For use with pages 396–401**

**TEST TAKING STRATEGY**  **Before you give up on a question, try to eliminate some of your choices so you can make an educated guess.**

1. *Multiple Choice*  Use the substitution method to determine the solution of the system of linear questions.

   $x + y = 8$   $y = 3x$

   Ⓐ  $(1, 7)$   Ⓑ  $(2, 6)$

   Ⓒ  $(-2, -6)$   Ⓓ  $(-1, -7)$

   Ⓔ  $(3, 5)$

2. *Multiple Choice*  Use the substitution method to determine the solution of the system of linear equations.

   $x + 2y = -6$   $2x - y = 8$

   Ⓐ  $(-2, -2)$   Ⓑ  $(-4, 2)$

   Ⓒ  $(2, -4)$   Ⓓ  $\left(-\frac{22}{5}, -\frac{26}{5}\right)$

   Ⓔ  $(-10, 1)$

3. *Multiple Choice*  Use the substitution method to determine the solution of the system of linear equations.

   $3x + 4y = 16$   $-3x + 2y = 8$

   Ⓐ  $(4, 0)$   Ⓑ  $(1, 4)$

   Ⓒ  $(-2, 5)$   Ⓓ  $(-2, 1)$

   Ⓔ  $(0, 4)$

4. *Multiple Choice*  Which point lies on the graph of the system?

   $4x + y = 23$   $x - y = 2$

   Ⓐ  $(3, 5)$   Ⓑ  $(5, 3)$

   Ⓒ  $(4, 7)$   Ⓓ  $(-2, 1)$

   Ⓔ  $(0, 4)$

5. *Multiple Choice*  Which point lies on the graph of the system?

   $y = x - 9$   $x + 2y = 0$

   Ⓐ  $(6, -3)$   Ⓑ  $(-3, 6)$

   Ⓒ  $(5, -4)$   Ⓓ  $(8, -1)$

   Ⓔ  $(6, 3)$

6. *Multiple Choice*  The ordered pair $(6, -8)$ is a solution of which system of equations?

   Ⓐ  $3x + 2y = 4$   Ⓑ  $x - 2y = 22$

   $x - y = 14$   $y = \frac{4}{5}x - 10$

   Ⓒ  $7x + 3y = 18$   Ⓓ  $2x + y = 4$

   $x - y = -2$   $-x - y = 2$

   Ⓔ  $y = -x - 2$

   $4x - 2y = 8$

7. *Multiple Choice*  If $5x - 3y = 1$ and $x - 2y = -4$, then $xy = $ __?__ .

   Ⓐ  5   Ⓑ  4   Ⓒ  6

   Ⓓ  8   Ⓔ  9

*Quantitative Comparison*  In Exercises 8–10, solve the linear system. Then choose the statement below that is true about the solution of the system.

   Ⓐ  The value of $x$ is greater than the value of $y$.

   Ⓑ  The value of $y$ is greater than the value of $x$.

   Ⓒ  The values of $x$ and $y$ are equal.

   Ⓓ  The relationship cannot be determined from the given information.

8.  $3x + y = 8$

    $-4x - 3y = -9$

9.  $5x + 2y = 28$

    $y = \frac{1}{2}x + 2$

10. $3y = 2x + 1$

    $7x - y = -13$

NAME _____ DATE _____

# *Standardized Test Practice*

For use with pages 402–408

**TEST TAKING STRATEGY   Think positively during a test. This will help keep up your confidence and enable you to focus on each question.**

1. *Multiple Choice*   Order the steps below to solve the system of equations using linear combinations.

   $2x + 5y = 7$        Equation 1

   $4y - 3x = 16$        Equation 2

   **I.**   Substitute the known variable into either of the original equations. Solve for the remaining unknown variable.

   **II.**   Multiply Equation 1 by 3 and Equation 2 by 2.

   **III.**   Multiply Equation 1 by 3 and Equation 2 by $-2$.

   **IV.**   Arrange the equations with like terms in columns.

   **V.**   Add the equations, combine like terms to eliminate one variable, and solve for the remaining variable.

   (A) I, II, III, IV        (B) IV, II, V, I

   (C) II, IV, I, V        (D) IV, III, V, I

   (E) III, IV, I, V

2. *Multiple Choice*   Solve the system of linear equations using linear combinations.

   $5x + 3y = 26$

   $-2x + 3y = -2$

   (A) $(6, -1)$        (B) $(-4, -2)$   (C) $(2, 5)$

   (D) $(4, 2)$        (E) $(7, -3)$

3. *Multiple Choice*   If $3x + 2y = 8$ and $-4x + 3y = -5$, then $x + y = $ ___?___ .

   (A) 6        (B) 8        (C) 7

   (D) 5        (E) 3

4. *Multiple Choice*   If $2y - 3 = 5x$ and $y = \frac{2}{3}x + 7$, then $xy = $ ___?___ .

   (A) 12        (B) 25        (C) 27

   (D) 18        (E) 14

5. *Multiple Choice*   Two hunters start from different points in the woods and hike towards their campsite. The first hunter travels a trail along the line $2y - x = 24$. The second hunter's trail follows the line $3y = -2x - 18$. They meet at the campsite. What are the coordinates of the site?

   (A) $\left(-\frac{108}{7}, \frac{30}{7}\right)$        (B) $\left(-\frac{30}{7}, \frac{108}{7}\right)$

   (C) $(-13, 2)$        (D) $(-14, 5)$

   (E) $\left(-\frac{120}{7}, \frac{48}{7}\right)$

6. *Multi-Step Problem*   Three oranges and 2 apples cost $1.70. Two oranges and 3 apples cost $1.55.

   **a.** Let $x$ represent the cost in dollars of an orange and let $y$ represent the cost in dollars of an apple. Write a system of linear equations to represent the situation.

   **b.** Solve the system.

   **c.** What is the cost of an apple?

Chapter 7

NAME _____ DATE _____

# *Standardized Test Practice*

**For use with pages 409–414**

**TEST TAKING STRATEGY   Go back and check as much of your work as you can.**

1. *Multiple Choice*   Your class is selling wrapping paper and candles for a fund raiser. Rolls of wrapping paper sell for $5 and candles sell for $4 each. A total of 526 items were sold and $2327 was raised. How many of each item was sold?

   Ⓐ   220 rolls, 306 candles

   Ⓑ   223 rolls, 303 candles

   Ⓒ   326 rolls, 200 candles

   Ⓓ   200 rolls, 326 candles

   Ⓔ   215 rolls, 311 candles

2. *Multiple Choice*   The manual for your boat engine calls for 91 octane gas. The gas stations by your house sell only 87 and 93 octane. If the boat's tank holds 30 gallons, how many gallons of each should you buy for a 91 octane mix?

   Ⓐ   15 gallons of 87 and 15 gallons of 93

   Ⓑ   12 gallons of 87 and 18 gallons of 93

   Ⓒ   10 gallons of 87 and 20 gallons of 93

   Ⓓ   20 gallons of 87 and 10 gallons of 93

   Ⓔ   18 gallons of 87 and 12 gallons of 93

3. *Multiple Choice*   You went skiing for 6 h. You skied at an average rate of 30 mi/h and the ski lift took you back up the hill at a rate of 10 mi/h. If your overall average rate was 23 mi/h, how many hours did you actually spend skiing? (*Hint*: Use the model: skiing speed · skiing time + ski lift speed · ski lift time = total distance.)

   Ⓐ   4        Ⓑ   3.5        Ⓒ   3.9

   Ⓓ   3.1      Ⓔ   4.5

4. *Multiple Choice*   You are taking a test worth 130 points. There are a total of 50 five-point and two-point questions. How many five-point questions are on the test?

   Ⓐ   15       Ⓑ   5         Ⓒ   20

   Ⓓ   12       Ⓔ   10

5. *Multiple Choice*   At what point do the lines $12x - 3y = 9$ and $y = -\frac{3}{2}x + 8$ intersect?

   Ⓐ   $(8, 9)$        Ⓑ   $(1, 4)$

   Ⓒ   $\left(\frac{30}{33}, \frac{21}{33}\right)$   Ⓓ   $(2, 5)$      Ⓔ   $(2, 4)$

6. *Multiple Choice*   A school has two soccer seasons. There are currently 30 students playing on the spring team and participation is increasing by 2 students per year. There are currently 19 students playing on the fall team and this is increasing by 3 students per year. When will the teams be the same size?

   Ⓐ   12 years     Ⓑ   14 years

   Ⓒ   10 years     Ⓓ   11 years

   Ⓔ   8 years

*Quantitative Comparison*   In Exercises 7–9, solve the linear system. Then choose the statement below that is true about the solution of the system.

   Ⓐ   The value of *x* is greater than the value of *y*.

   Ⓑ   The value of *y* is greater than the value of *x*.

   Ⓒ   The values of *x* and *y* are equal.

   Ⓓ   The relationship cannot be determined from the given information.

7. $2x + 5y = -16$
   $4x + 3y = -4$

8. $3y = 10x - 14$
   $y = \frac{3}{2}x - 1$

9. $8y - 3x = 31$
   $3x + 5y = 34$

**Algebra 1, Concepts and Skills**
Standardized Test Practice Workbook

# Standardized Test Practice

**For use with pages 417–422**

**TEST TAKING STRATEGY** Avoid spending too much time on one question. Skip questions that are too difficult for you and spend no more than a few minutes on each question.

1. **Multiple Choice** How many solutions does the linear system have?

$$y = \frac{3}{2}x + 8$$
$$6x - 4y = 16$$

(A) None (B) Exactly one
(C) Two (D) Infinitely many
(E) Cannot be determined

2. **Multiple Choice** How many solutions does the linear system have?

$$x = 3$$
$$y = -2$$

(A) None (B) Exactly one
(C) Two (D) Infinitely many
(E) Cannot be determined

3. **Multiple Choice** How many solutions does the linear system have?

$$6x - 2y = 14$$
$$y = 3x - 9$$

(A) None (B) Exactly one
(C) Two (D) Infinitely many
(E) Cannot be determined

4. **Multiple Choice** What is the solution of the system of linear equations?

$$y = \frac{1}{2}x + 8$$
$$2x + y = 18$$

(A) $(-4, -10)$ (B) $(4, 10)$ (C) $(2, 5)$
(D) No solution (E) Cannot be determined

5. **Multiple Choice** What is the solution of the system of linear equations?

$$5x - 2y = 16$$
$$y = \frac{5}{2}x - 10$$

(A) $(2, -3)$ (B) $(4, 2)$
(C) $(0, -8)$ (D) No solution
(E) Infinitely many

6. **Multiple Choice** How many solutions does the graph below have?

(A) None
(B) Exactly one
(C) Two
(D) Infinitely many
(E) Cannot be determined

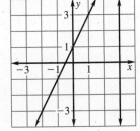

**Quantitative Comparison** In Exercises 7–9, solve the linear system. Then choose the statement below that is true about the solution of the system.

(A) The number of solutions in column A is greater.
(B) The number of solutions in column B is greater.
(C) The number of solutions are equal.
(D) The relationship cannot be determined from the given information.

| | Column A | Column B |
|---|---|---|
| 7. | $y = \frac{1}{2}x + 2$ <br> $-x + 2y = 4$ | $x + y = 7$ <br> $3x + 4y = 16$ |
| 8. | $y = 3x + 6$ <br> $y = 3x - 2$ | $-3x + 2y = 2$ <br> $y = \frac{3}{2}x + 6$ |
| 9. | $5x = 10$ <br> $y = -7$ | $7x + 2y = 12$ <br> $14x - 24 = -4y$ |

NAME _____ DATE _____

# Standardized Test Practice

For use with pages 424–430

**TEST TAKING STRATEGY**  **Read all of the answer choices before deciding which is the correct one.**

1. **Multiple Choice**  Choose the system of linear inequalities shown in the graph.

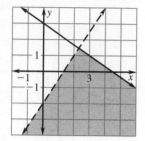

   **A**  $y < \frac{3}{2}x - 2$     **B**  $y > \frac{3}{2}x - 2$
   $y \le -\frac{2}{3}x + 3$         $y \ge -\frac{2}{3}x + 3$

   **C**  $y > \frac{3}{2}x - 2$     **D**  $y < \frac{3}{2}x - 2$
   $y \le -\frac{2}{3}x + 3$         $y \ge -\frac{2}{3}x + 3$

   **E**  None of these

2. **Multiple Choice**  Choose the system of linear inequalities shown in the graph.

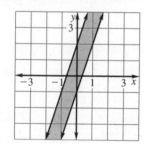

   **A**  $y \le 3x + 2$     **B**  $y \ge 3x + 2$
   $y \le 3x - 1$            $y \le 3x - 1$

   **C**  $y \ge \frac{1}{3}x + 2$     **D**  $y \le 3x + 2$
   $y \ge \frac{1}{3}x - 1$            $y \ge 3x - 1$

   **E**  None of these

3. **Multiple Choice**  Which point is a solution of the system of linear inequalities?

   $y \le \frac{1}{2}x + 2$
   $y < -\frac{2}{3}x - 1$

   **A**  $(0, 0)$     **B**  $(-2, -3)$   **C**  $(1, 2)$
   **D**  $(-2, 3)$    **E**  $(-4, 2)$

4. **Multiple Choice**  Choose the system of linear inequalities shown in the graph.

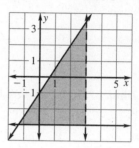

   **A**  $x > 3$           **B**  $x < 3$
   $y \ge -3$               $y \le -3$
   $y \le \frac{3}{2}x - 1$   $y \le \frac{3}{2}x - 1$

   **C**  $x < 3$           **D**  $x < 3$
   $y \le -3$               $y \ge -3$
   $y \ge \frac{3}{2}x - 1$   $y \le \frac{3}{2}x - 1$

   **E**  None of these

5. **Multiple Choice**  Which point is a solution of the system of linear inequalities?

   $y < x + 1$
   $y \ge 2x$

   **A**  $(1, -3)$     **B**  $(0, 3)$     **C**  $(2, 0)$
   **D**  $(-3, -8)$    **E**  $(-2, -2)$

6. **Multi-Step Problem**  During the summer you take two part-time jobs. The first pays $5 an hour. The second pays $8 an hour. You want to earn at least $150 a week and work 25 hours or less a week.

   **a.**  Write a system of inequalities that model the hours you can work at each job.

   **b.**  Graph the system.

   **c.**  List two alternatives for dividing your time between jobs.

Chapter 7

# *Standardized Test Practice*

For use with pages 443–448

**TEST TAKING STRATEGY**   **Be aware of how much time you have left, but keep focused on your work.**

**1. *Multiple Choice***   How many factors are in the expression $5^3 \cdot 5^6$?

Ⓐ  5          Ⓑ  9          Ⓒ  3

Ⓓ  2          Ⓔ  1

**2. *Multiple Choice***   In the expression $x^7$, the 7 is called the ____?____.

Ⓐ  factor          Ⓑ  base

Ⓒ  exponent          Ⓓ  product

Ⓔ  term

**3. *Multiple Choice***   Simplify $4^3 \cdot 4^5$. Write the answer as a power.

Ⓐ  $8^8$          Ⓑ  $16^8$          Ⓒ  $4^{15}$

Ⓓ  $4^8$          Ⓔ  $16^{15}$

**4. *Multiple Choice***   Simplify $(5^2)^7$. Write the answer as a power.

Ⓐ  $25^9$          Ⓑ  $25^{14}$          Ⓒ  $5^5$

Ⓓ  $5^9$          Ⓔ  $5^{14}$

**5. *Multiple Choice***   Simplify $(-3xy^2)^4$. Write the answer as a power.

Ⓐ  $81x^4y^8$          Ⓑ  $-81x^4y^8$

Ⓒ  $12x^4y^8$          Ⓓ  $-12x^5y^8$

Ⓔ  $81x^5y^6$

**6. *Multiple Choice***   Evaluate $(a^2 \cdot a^4)^2$ when $a = 2$.

Ⓐ  256          Ⓑ  4096          Ⓒ  16

Ⓓ  144          Ⓔ  128

**7. *Multiple Choice***   Evaluate $-(x^3)^2(3x^2)$ when $x = -3$.

Ⓐ  19,683          Ⓑ  6561

Ⓒ  $-19,683$          Ⓓ  $-6561$

Ⓔ  $-59,049$

**8. *Multiple Choice***   The volume $V$ of a cylinder  is given by $V = \pi r^2 h$, where $r$ is the radius, $h$ is the height and $\pi$ is approximately 3.14. What is the volume of the cylinder in terms of $x$?

Ⓐ  $94.2x^2$

Ⓑ  $282.6x^2$

Ⓒ  $94.2x^4$

Ⓓ  $282.6x^4$

Ⓔ  $887.4x^4$

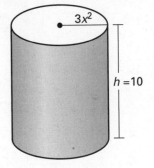

*Quantitative Comparison*   In Exercises 9–12, simplify each expression. Then choose the statement below that is true about the given numbers.

Ⓐ   The number in column A is greater.

Ⓑ   The number column B is greater.

Ⓒ   The two numbers are equal.

Ⓓ   The relationship cannot be determined with the given information.

|   | *Column A* | *Column B* |
|---|---|---|
| **9.** | $x^7 \cdot x^5$ | $(x^7)^5$ |
| **10.** | $(2x^2)^3$ | $(4x^2)(2x^4)$ |
| **11.** | $(3x)^2(x^3)$ when $x = -3$ | $(7x^2)^3$ when $x = -2$ |
| **12.** | $(ab^2)(x^3)$ when $a = 2$ and $b = -1$ | $(b^2)^3(ab)$ when $a = 1$ and $b = -2$ |

**Algebra 1, Concepts and Skills**
Standardized Test Practice Workbook

# Standardized Test Practice

For use with pages 449–454

**TEST TAKING STRATEGY** Learn as much as you can about the test ahead of time, such as types of questions and the topics the test will cover.

**1. Multiple Choice** Evaluate $(5^{-3})\left(\dfrac{1}{5^{-5}}\right)(5^{-4})$.

- (A) 10
- (B) $-25$
- (C) $\dfrac{1}{25}$
- (D) $\dfrac{1}{10}$
- (E) $-\dfrac{1}{25}$

**2. Multiple Choice** Evaluate $2^{-2} \cdot 2^4 \cdot 2^{-3}$.

- (A) $\dfrac{1}{4}$
- (B) $\dfrac{1}{2}$
- (C) 2
- (D) 4
- (E) $-2$

**3. Multiple Choice** Evaluate $(3^3)^{-4}(3^7)$.

- (A) 243
- (B) $\dfrac{1}{243}$
- (C) $-243$
- (D) 6561
- (E) $\dfrac{1}{6561}$

**4. Multiple Choice** Rewrite $(3x^{-2}y)(6xy^{-3})$ with positive exponents.

- (A) $\dfrac{18}{xy^2}$
- (B) $\dfrac{18x}{y^2}$
- (C) $\dfrac{9}{xy^2}$
- (D) $9xy^2$
- (E) $\dfrac{18}{x^3y^4}$

**5. Multiple Choice** Rewrite $(2x^{-2}y)(3y^{-2})$ with positive exponents.

- (A) $\dfrac{6y}{x^2}$
- (B) $\dfrac{5y}{x^2}$
- (C) $\dfrac{6y^{-3}}{x^2}$
- (D) $\dfrac{6y^3}{x^2}$
- (E) $\dfrac{6}{x^2y}$

**6. Multiple Choice** Evaluate $(-3)^2(-3)^{-2}$.

- (A) 81
- (B) 1
- (C) $-1$
- (D) 0
- (E) $-81$

**7. Multiple Choice** Rewrite $(-3x^3y^{-6})(2x^{-3}y)$ with positive exponents.

- (A) $\dfrac{-6x}{y^7}$
- (B) $\dfrac{-6}{xy^5}$
- (C) $\dfrac{-6x}{y^5}$
- (D) $\dfrac{-6}{y^5}$
- (E) $\dfrac{-5}{y^5}$

**8. Multiple Choice** Rewrite $-\dfrac{2x^{-1}}{x^2y^{-1}}$ with positive exponents.

- (A) $-2x^3y$
- (B) $\dfrac{2x^3}{y}$
- (C) $-\dfrac{2y}{x^3}$
- (D) $-\dfrac{2y}{x}$
- (E) $\dfrac{2y}{x}$

**9. Multiple Choice** Evaluate $\dfrac{2}{2^{-2}}$.

- (A) 8
- (B) 4
- (C) $-4$
- (D) $\dfrac{1}{8}$
- (E) $-8$

**Quantitative Comparison** In Exercises 10–12, choose the statement below that is true about the given values.

- (A) The value in column A is greater.
- (B) The value in column B is greater.
- (C) The two values are equal.
- (D) The relationship cannot be determined with the given information.

| | Column A | Column B |
|---|---|---|
| **10.** | $2^0$ | $4^0$ |
| **11.** | $4^{-2}$ | $4^{-3}$ |
| **12.** | $\dfrac{1}{3^{-2}}$ | $\dfrac{1}{4^{-2}}$ |

**Chapter 8**

NAME _____ DATE _____

## *Standardized Test Practice*

**For use with pages 455–461**

**TEST TAKING STRATEGY** **Learn as much as you can about the test ahead of time, such as types of questions and the topics the test will cover.**

1. *Multiple Choice* Which function contains the point (0, 1)?

   **A** $y = 3^x$        **B** $y = 2(3)^x$

   **C** $y = 2\left(\frac{1}{3}\right)^x$   **D** $y = 3\left(\frac{1}{3}\right)^x$

   **E** None of these

2. *Multiple Choice* Evaluate $y = 2(5)^x$ when $x = 2.2$. Round the answer to the nearest hundredth.

   **A** 66.9        **B** 66.99

   **C** 69.89       **D** 68.99

   **E** 68.9

3. *Multiple Choice* Describe the domain and range of the function $y = -\left(\frac{1}{2}\right)^x$.

   **A** domain: all real numbers; range: all positive real numbers

   **B** domain: all negative real numbers; range: all negative real numbers

   **C** domain: all real numbers; range: all negative real numbers

   **D** domain: all positive real numbers; range: all positive real numbers

   **E** domain: all real numbers; range: all real numbers

4. *Multiple Choice* What point do the following functions have in common?

   $$y = \frac{1}{2}(3)^x \qquad y = \frac{4^x}{2}$$

   **A** $\left(0, \frac{1}{2}\right)$   **B** $\left(\frac{1}{2}, 0\right)$

   **C** $(2, 0)$       **D** $(0, 2)$

   **E** $(0, 0)$

5. *Multiple Choice* Choose the equation of the curve shown.

   **A** $y = (-1)^x$
   **B** $y = 2^x$
   **C** $y = 3^x$
   **D** $y = \left(\frac{1}{2}\right)^x$
   **E** $y = \left(\frac{1}{3}\right)^x$

   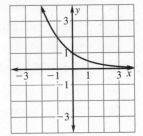

6. *Multiple Choice* Choose the equation of the curve shown.

   **A** $y = 3^{-x}$
   **B** $y = 5^{-x}$
   **C** $y = \left(\frac{1}{3}\right)^{-x}$
   **D** $y = \left(\frac{1}{5}\right)^{-x}$
   **E** $y = (-5)^x$

   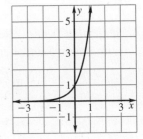

7. *Multi-Step Problem* You started a savings account in 1996. The balance $y$ is modeled by the equation $y = 500(1.08)^t$, where $t = 0$ represents the year 2000.

   **a.** What is the balance in 1996?

   **b.** What is the balance in 2000?

   **c.** What is the balance in 2002?

**Algebra 1, Concepts and Skills**
Standardized Test Practice Workbook

Chapter 8

NAME _____ DATE _____

# Standardized Test Practice

**For use with pages 462–468**

**TEST TAKING STRATEGY**  **Before you give up on a question, try to eliminate some of your choices so you can make an educated guess.**

**1. Multiple Choice**  Evaluate $\dfrac{3^7}{3^4}$.

  **A** $\frac{1}{27}$     **B** $27$     **C** $9$

  **D** $\frac{1}{9}$     **E** $-27$

**2. Multiple Choice**  Evaluate $\dfrac{-3^5}{(-3)^5}$.

  **A** $-1$     **B** $0$     **C** $3$

  **D** $-3$     **E** $1$

**3. Multiple Choice**  Evaluate $\left(-\dfrac{2}{5}\right)^2$.

  **A** $-\frac{25}{4}$     **B** $\frac{4}{10}$     **C** $\frac{4}{25}$

  **D** $-\frac{4}{25}$     **E** $\frac{4}{5}$

**4. Multiple Choice**  Evaluate $\left(\dfrac{3}{4}\right)^{-3}$.

  **A** $-\frac{9}{12}$     **B** $\frac{12}{9}$     **C** $\frac{27}{64}$

  **D** $\frac{64}{27}$     **E** $-\frac{27}{64}$

**5. Multiple Choice**  Simplify $\dfrac{3x^5y}{2xy^2} \cdot \dfrac{6xy^4}{4x^7}$.

  **A** $\frac{9y^3}{4x^2}$     **B** $\frac{9y^3}{8x^2}$     **C** $\frac{9y^2}{4x}$

  **D** $\frac{9x^2y^2}{8}$     **E** $\frac{9}{4}x^2y^3$

**6. Multiple Choice**  Simplify $\dfrac{5xy^0}{2y^3} \cdot \dfrac{12x^2y^7}{5x}$.

  **A** $6x^2y^4$     **B** $6x^4y^5$     **C** $\frac{6x^4}{y^4}$

  **D** $6xy^4$     **E** $6y^4$

**7. Multiple Choice**  Which expression simplifies to $x^2$?

  **A** $\frac{x^5}{x^{-3}}$     **B** $\frac{5x^5y}{5x^{-4}y}$     **C** $\frac{3x^{-3}}{3x^5}$

  **D** $\frac{x^{-7}y^2}{x^{-9}y^2}$     **E** $\frac{x^{-9}y^2}{x^{-7}y^2}$

**8. Multiple Choice**  Simplify $\left(\dfrac{5x^3y}{3y^2}\right)^2 \cdot \left(\dfrac{3^2xy^3}{5x^8}\right)$.

  **A** $\frac{5}{x}$     **B** $5x$     **C** $5xy$

  **D** $\frac{5y}{x}$     **E** $\frac{5y^2}{x}$

**9. Multiple Choice**  Simplify $\dfrac{(x^2y)^3}{2xy^2} \cdot \left(\dfrac{3x^4y^2}{2xy}\right)^{-2}$.

  **A** $\frac{9x^2}{8y}$     **B** $\frac{3}{4}x^{10}y$     **C** $\frac{2}{9}xy$

  **D** $\frac{2}{9}x^2y$     **E** $\frac{2}{9xy}$

**10. Multiple Choice**  Simplify $\left(\dfrac{xy^2z}{yz^4}\right)^{-2} \cdot \dfrac{y}{x^{-2}}$.

  **A** $yz^6$     **B** $y^2z^6$     **C** $\frac{z^6}{y^2}$

  **D** $\frac{z^6}{y}$     **E** $\frac{y^2}{z^6}$

**Quantitative Comparison**  In Exercises 11–13, simplify the expression using $x = -1$ and $y = 2$. Then choose the statement below that is true about the given numbers.

  **A**  The number in column A is greater.

  **B**  The number in column B is greater.

  **C**  The two numbers are equal.

  **D**  The relationship cannot be determined from the given information.

| | Column A | Column B |
|---|---|---|
| **11.** | $x^y$ | $y^x$ |
| **12.** | $(5^x)(5yx^2)$ | $\left(\frac{1}{2}\right)^x$ |
| **13.** | $(3y^3x^{-y})(6y^x)$ | $(5^yx^3) \cdot x^y$ |

Chapter 8

**TEST TAKING STRATEGY** **As soon as the testing begins, start working. Keep moving and stay focused on the test.**

1. *Multiple Choice* Rewrite $5.4 \times 10^4$ in decimal form.

   Ⓐ 54,000          Ⓑ 540,000

   Ⓒ 0.000054        Ⓓ 0.00054

   Ⓔ 5400

2. *Multiple Choice* Rewrite $8.2 \times 10^{-6}$ in decimal form.

   Ⓐ 0.00000082      Ⓑ 0.000082

   Ⓒ 8,200,000       Ⓓ 0.0000082

   Ⓔ 820,000

3. *Multiple Choice* Rewrite 0.00000036 in scientific notation.

   Ⓐ $3.6 \times 10^7$    Ⓑ $3.6 \times 10^{-7}$

   Ⓒ $3.6 \times 10^6$    Ⓓ $3.6 \times 10^{-6}$

   Ⓔ $3.6 \times 10^{-8}$

4. *Multiple Choice* Which of the following numbers is *not* written in scientific notation?

   Ⓐ $1.2 \times 10^8$      Ⓑ $3.72 \times 10^{-7}$

   Ⓒ $8 \times 10^5$        Ⓓ $76.02 \times 10^9$

   Ⓔ $1.408 \times 10^{-2}$

5. *Multiple Choice* Evaluate the product $(7.2 \times 10^5) \cdot (6.1 \times 10^{-8})$. Write the result in scientific notation.

   Ⓐ $43.92 \times 10^{-2}$   Ⓑ $4.392 \times 10^{-3}$

   Ⓒ $4.392 \times 10^3$      Ⓓ $4.392 \times 10^2$

   Ⓔ $4.392 \times 10^{-2}$

6. *Multiple Choice* Evaluate $(3.0 \times 10^{-3})^4$.

   Ⓐ $8.1 \times 10^{-7}$    Ⓑ $9 \times 10^{-12}$

   Ⓒ $8.1 \times 10^{-11}$   Ⓓ $8.1 \times 10^{-12}$

   Ⓔ $8.1 \times 10^{-13}$

7. *Multiple Choice* Evaluate $\dfrac{9.6 \times 10^{-4}}{1.2 \times 10^{-6}}$.

   Ⓐ $7 \times 10^{10}$    Ⓑ $8 \times 10^2$

   Ⓒ $8.4 \times 10^2$    Ⓓ $8.4 \times 10^{-2}$

   Ⓔ $7 \times 10^{-2}$

8. *Multiple Choice* An astronomer measured the speed of a comet at 150,000 miles per hour. Write the number in scientific notation.

   Ⓐ $1.5 \times 10^6$    Ⓑ $1.5 \times 10^{-6}$

   Ⓒ $1.5 \times 10^5$    Ⓓ $1.5 \times 10^{-5}$

   Ⓔ $1.5 \times 10^{-4}$

9. *Multiple Choice* Evaluate $\dfrac{(0.002)^3}{400,000}$. Write the result in scientific notation.

   Ⓐ $2 \times 10^{-14}$    Ⓑ $2 \times 10^{-4}$

   Ⓒ $1 \times 10^{-14}$    Ⓓ $1 \times 10^{-4}$

   Ⓔ $2 \times 10^{14}$

*Quantitative Comparison* In Exercises 10–12, choose the statement below that is true about the given values.

   Ⓐ The value in column A is greater.

   Ⓑ The value column B is greater.

   Ⓒ The two values are equal.

   Ⓓ The relationship cannot be determined with the given information.

| | Column A | Column B |
|---|---|---|
| 10. | $8.6 \times 10^6$ | $8.6 \times 10^{-6}$ |
| 11. | $(2.3 \times 10^{-5})^2$ | $\dfrac{4.84 \times 10^{-4}}{8.8 \times 10^3}$ |
| 12. | $(3.2 \times 10^5)(8.6 \times 10^7)$ | $\dfrac{(2 \times 10^5)^2}{1 \times 10^{-3}}$ |

NAME _____ DATE _____

# *Standardized Test Practice*

**For use with pages 476–481**

**TEST TAKING STRATEGY**   **Think positively during a test. This will keep up your confidence and enable you to focus on each question.**

1. *Multiple Choice*   In the model
   $y = C(1 + r)^t$, $(1 + r)$ is the ___?___.
   **(A)** time period       **(B)** initial amount
   **(C)** growth factor     **(D)** growth rate
   **(E)** percent increase

2. *Multiple Choice*   You deposit $1500 into a savings account that pays 6% interest compounded yearly. How much money is in the account after 10 years assuming you made no additional deposits or withdrawals?
   **(A)** $1590        **(B)** $2400
   **(C)** $2657.34     **(D)** $2686.27
   **(E)** $3890.61

3. *Multiple Choice*   You deposit $750 into a savings account that pays 8% interest compounded yearly. How much money is in the account after 7 years assuming no additional deposits or withdrawals were made?
   **(A)** $1170        **(B)** $1285.37
   **(C)** $1204.34     **(D)** $1388.20
   **(E)** $1560

4. *Multiple Choice*   Which model best represents the growth curve shown in the graph?
   **(A)** $y = 25(3)^t$
   **(B)** $y = 25(1.5)^t$
   **(C)** $y = 25(0.9)^t$
   **(D)** $y = 50(1.1)^t$
   **(E)** $y = 50(0.9)^t$

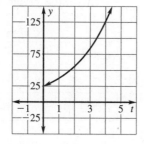

5. *Multiple Choice*   A wildlife management group releases 12 elk in a reintroduction program area. The population is expected to increase by 25% each year for the next 5 years. What is the estimated elk population after 5 years? (Round down to the nearest whole number.)
   **(A)** 27       **(B)** 37       **(C)** 36
   **(D)** 15       **(E)** 38

6. *Multiple Choice*   You start a new job at a pay rate of $6 per hour. You expect a raise of 4% each year. After 6 raises, how much will you be earning per hour?
   **(A)** $7.44     **(B)** $7.02     **(C)** $7.57
   **(D)** $7.59     **(E)** $8.51

7. *Multi-Step Problem*   You bought a piece of land for $50,000 in 1998 in an area of rapidly increasing population growth. You expect the land to increase in value 15% each year for the next 10 years.

   **a.** Write an exponential growth model for the situation.

   **b.** Estimate the value of the land in 2005. (Round to the nearest dollar.)

   **c.** Estimate the value of the land in 1996. (Round to the nearest dollar.)

   **d.** *Writing*   If the population growth begins to level off, what might happen to the value of the land?

*Chapter 8*

# Standardized Test Practice

For use with pages 482–488

**TEST TAKING STRATEGY** **Avoid spending too much time on one question. Skip questions that are too difficult for you, and spend no more than a few minutes on each question.**

1. *Multiple Choice*  In an exponential decay function, the decay factor is always ____?____ .

   A  greater than zero

   B  less than zero

   C  greater than one

   D  less than one

   E  A and D

2. *Multiple Choice*  Which model is an exponential decay model?

   A  $y = 5x + 6$    B  $y = 5(1.2)^t$

   C  $y = 7(0.9)^t$    D  $y = 5 - 3t$

   E  $y = 50 - 3(1.1)^t$

3. *Multiple Choice*  Which model best represents the decay curve shown in the graph?

   A  $y = 50(0.76)^{-t}$

   B  $y = 50(0.76)^t$

   C  $y = 100(0.76)^t$

   D  $y = 100(1.32)^t$

   E  $y = 100(0.76)^{-t}$

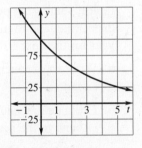

4. *Multiple Choice*  You buy a new laptop computer for $3500 in 1998. The computer depreciates at the rate of 18% per year. What is its value in 2001?

   A  $1929.79    B  $1890

   C  $1610    D  $4130

   E  $2870

5. *Multiple Choice*  You are in your state's high school tennis championship tournament. At the start of the event there are 128 participants, and each round eliminates half of the players. How many players remain after round 3?

   A  64    B  32    C  16

   D  8    E  4

6. *Multiple Choice*  You have a large oak tree in your yard. If the tree has 10,000 leaves at the beginning of autumn and loses 7% each day until all the leaves are off, how many leaves remain on the tree after 1 week of leaf loss?

   A  4900    B  9300

   C  6957    D  6017

   E  5000

*Quantitative Comparison*  In Exercises 7–9, evaluate each function. Then choose the statement below that is true about the given values of y.

   A  The value of y in column A is greater.

   B  The value of y column B is greater.

   C  The two values of y are equal.

   D  The relationship cannot be determined with the given information.

| | | Column A | Column B |
|---|---|---|---|
| 7. | $t = 1$ | $y = 26(0.8)^{-t}$ | $y = 26(1.2)^t$ |
| 8. | $t = -1$ | $y = 5.6\left(\frac{2}{3}\right)^t$ | $y = 5.6(1.5)^{-t}$ |
| 9. | $t = 2$ | $y = 14\left(\frac{1}{2}\right)^t$ | $y = 17\left(\frac{2}{5}\right)^t$ |

# Standardized Test Practice

**For use with pages 499–504**

**TEST TAKING STRATEGY**   Work as fast as you can through the easier problems, but not so fast that you are careless.

*Multiple Choice*   In Exercises 1–4, evaluate the expression, rounding to the nearest hundredth when necessary.

**1.** $\pm\sqrt{100}$

   (A)  12, −12  (B)  12    (C)  −12

   (D)  10, −10  (E)  10

**2.** $-\sqrt{37}$

   (A)  −6.1  (B)  6.1   (C)  −6

   (D)  6.08  (E)  −6.08

**3.** $\sqrt{324}$

   (A)  17.29  (B)  28   (C)  16

   (D)  18  (E)  16.33

**4.** $\pm\sqrt{19}$

   (A)  4.35, −4.35

   (B)  4.3, −4.3

   (C)  4.36

   (D)  4.36, −4.36

   (E)  4.4, −4.4

*Multiple Choice*   In Exercises 5–8, evaluate the expression, rounding to the nearest hundredth when necessary.

**5.** $\sqrt{b^2 - 4ac}$ when $a = 2$, $b = -3$, and $c = -9$.

   (A)  5    (B)  7    (C)  10.20

   (D)  8.31  (E)  9

**6.** $\sqrt{b^2 - 4ac}$ when $a = 4$, $b = 12$, and $c = 9$.

   (A)  0    (B)  10.39  (C)  21.17

   (D)  −10.39  (E)  16.97

**7.** $\frac{a\sqrt{b} - 4}{b}$ when $a = 9$ and $b = 16$.

   (A)  8.75  (B)  2    (C)  3.56

   (D)  2.75  (E)  3

**8.** $7 \pm 4\sqrt{2}$

   (A)  11, 3  (B)  15.56, −15.56

   (C)  13.48, 0.52  (D)  15.56, 4.24

   (E)  12.66, 1.34

**9.** *Multiple Choice*  Which of the following is an example of a perfect square?

   (A)  782  (B)  688  (C)  174

   (D)  482  (E)  576

*Quantitative Comparison*   In Exercises 10–12, choose the statement below that is true about the given numbers.

   (A)  The number in column A is greater.

   (B)  The number in column B is greater.

   (C)  The two numbers are equal.

   (D)  The relationship cannot be determined from the given information.

|  | *Column A* | *Column B* |
|---|---|---|
| **10.** | $\sqrt{2}$ | 1.41 |
| **11.** | $\sqrt{2^2}$ | $\sqrt{(-2)^2}$ |
| **12.** | $\sqrt{\dfrac{1}{4}}$ | $\sqrt{\dfrac{1}{5}}$ |

**Algebra 1, Concepts and Skills**
Standardized Test Practice Workbook

*Chapter 9*

# *Standardized Test Practice*
**For use with pages 505–510**

**TEST TAKING STRATEGY** **Work as fast as you can through the easier problems, but not so fast that you are careless.**

1. *Multiple Choice* Which quadratic equation is written in standard form?

Ⓐ $x^2 - 9 + x = 0$

Ⓑ $3x^2 + 5x + 1 = 0$

Ⓒ $7 - 4x^2 + 9 = 0$

Ⓓ $32 - 2x^2 = 0$

Ⓔ $2x - x^2 = 0$

2. *Multiple Choice* Consider the equation $x^2 = -4$. Which statement is correct?

Ⓐ The equation has exactly one solution.

Ⓑ The equation has two solutions.

Ⓒ The equation has three solutions.

Ⓓ The equation has no real solution.

Ⓔ The number of solutions cannot be determined.

3. *Multiple Choice* Solve $9x^2 - 81 = 0$.

Ⓐ 9     Ⓑ ± 9     Ⓒ ±3

Ⓓ 3     Ⓔ −3

4. *Multiple Choice* Solve $2t^2 + 3 = 35$.

Ⓐ ±8     Ⓑ ±2     Ⓒ ±16

Ⓓ ±4     Ⓔ 16

5. *Multiple Choice* An object is dropped from a height of 350 feet. To the nearest hundredth of a second, about how long does it take the object to hit the ground? Assume there is no air resistance.

Ⓐ 5.92 sec     Ⓑ 18.28 sec

Ⓒ 4.68 sec     Ⓓ 1.91 sec

Ⓔ 4.83 sec

6. *Multiple Choice* The sales $S$ (in dollars) of camping equipment at a small store can be modeled by $S = 52.6t^2 + 6500$, where $t$ is the number of years since 1990. Estimate the year in which the store's sales of camping equipment will be $12,800.

Ⓐ 2001     Ⓑ 2005     Ⓒ 1998

Ⓓ 1999     Ⓔ 2003

*Quantitative Comparison* In Exercises 7–10, solve the quadratic equation. Then choose the statement that is true about the positive value of $x$ in each solution.

Ⓐ The positive value of $x$ in column A is greater.

Ⓑ The positive value of $x$ in column B is greater.

Ⓒ The positive values of $x$ are equal.

Ⓓ The relationship cannot be determined from the given information.

|     | Column A | Column B |
| --- | --- | --- |
| 7.  | $5x^2 - 125 = 0$ | $7x^2 = 175$ |
| 8.  | $2x^2 - 72 = 0$ | $2x^2 = 128$ |
| 9.  | $3x^2 - 25 = 2$ | $2x^2 - 7 = 91$ |
| 10. | $5x^2 - 18 = 0$ | $3x^2 - 17 = 23$ |

**Algebra 1, Concepts and Skills**
Standardized Test Practice Workbook

NAME _____ DATE _____

# Standardized Test Practice

**For use with pages 511–517**

**TEST TAKING STRATEGY** Some questions involve more than one step. Reading too quickly might lead to mistaking the answer to a preliminary step for your final answer.

1. *Multiple Choice* Which one of the following is the simplified form of $\sqrt{192}$?

  **A** $3\sqrt{4}$    **B** $3\sqrt{8}$    **C** $4\sqrt{3}$

  **D** $8\sqrt{3}$    **E** $4\sqrt{12}$

2. *Multiple Choice* Which one of the following is the simplified form of $\frac{1}{3}\sqrt{450}$?

  **A** $\frac{5}{3}\sqrt{18}$    **B** $\frac{2}{3}\sqrt{15}$    **C** $5\sqrt{2}$

  **D** $12\sqrt{2}$    **E** $3\sqrt{50}$

3. *Multiple Choice* Which one of the following is the simplified form of $-5\sqrt{\frac{20}{3}}$?

  **A** $-15\sqrt{15}$    **B** $-5\sqrt{\frac{20}{3}}$

  **C** $-\frac{5}{3}\sqrt{20}$    **D** $-15\sqrt{60}$

  **E** $\frac{-10\sqrt{15}}{3}$

4. *Multiple Choice* Which one of the following is the simplified form of $\frac{1}{3}\sqrt{108}$?

  **A** $2\sqrt{3}$    **B** $\frac{2\sqrt{27}}{3}$    **C** $6$

  **D** $6\sqrt{3}$    **E** $\sqrt{12}$

5. *Multiple Choice* Simplify $\sqrt{\frac{25}{7}}$.

  **A** $\frac{\sqrt{175}}{7}$    **B** $7\sqrt{5}$    **C** $\frac{5}{\sqrt{7}}$

  **D** $\frac{5\sqrt{7}}{7}$    **E** $\frac{7\sqrt{5}}{5}$

6. *Multiple Choice* Find the area of the triangle using the formula $A = \frac{1}{2}bh$.

  **A** $3\sqrt{15}$

  **B** $15\sqrt{3}$

  **C** $\frac{13}{2}\sqrt{2}$

  **D** $5\sqrt{3}$

  **E** $3\sqrt{5}$

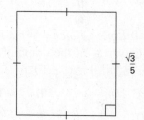

7. *Multiple Choice* Find the area of a square whose side measures $\frac{\sqrt{3}}{5}$.

  **A** $\frac{1}{25}\sqrt{3}$

  **B** $25\sqrt{3}$

  **C** $9\sqrt{5}$

  **D** $\frac{3}{25}$

  **E** $\frac{9}{25}$

*Quantitative Comparison* In Exercises 8–11, perform the indicated operation and simplify the result. Then choose the statement below that is true about the given numbers.

  **A** The number in column A is greater.

  **B** The number in column B is greater.

  **C** The numbers are equal.

  **D** The relationship cannot be determined from the given information.

| | Column A | Column B |
|---|---|---|
| 8. | $3\sqrt{8}$ | $\sqrt{\frac{72}{4}}$ |
| 9. | $4\sqrt{\frac{5}{4}}$ | $18\sqrt{\frac{2}{9}}$ |
| 10. | $\frac{9}{5}\sqrt{100}$ | $\frac{6}{5}\sqrt{225}$ |
| 11. | $-2\sqrt{72}$ | $-\sqrt{216}$ |

**Algebra 1, Concepts and Skills**
Standardized Test Practice Workbook

*Chapter 9*

**TEST TAKING STRATEGY**   Be aware of how much time you have left, but keep focused on your work.

**1.** *Multiple Choice*   The graph of $y = ax^2 + bx + c$ is a parabola whose vertex has an *x*-coordinate of ___?___ .

(A) $-\dfrac{2a}{b}$     (B) $\dfrac{b}{2a}$     (C) $-\dfrac{a}{2b}$

(D) $-\dfrac{b}{2a}$     (E) $\dfrac{a}{2b}$

**2.** *Multiple Choice*   What is the *x*-coordinate of the vertex of the graph of the equation $y = -x^2 + 4x - 6$?

(A) $-2$     (B) $2$     (C) $-\frac{1}{8}$

(D) $\frac{1}{8}$     (E) $0$

**3.** *Multiple Choice*   What is the *x*-coordinate of the vertex of the graph of the equation $y = \frac{1}{4}x^2 + 5x - 10$?

(A) $0$     (B) $-10$     (C) $10$

(D) $-2$     (E) $-\frac{5}{4}$

**4.** *Multiple Choice*   What is the equation of the axis of symmetry for the equation $y = 6x^2 - 4x + 3$?

(A) $y = \frac{1}{3}$     (B) $y = 3$     (C) $x = \frac{1}{3}$

(D) $x = 3$     (E) $x = \frac{3}{4}$

**5.** *Multiple Choice*   What is the *y*-coordinate of the vertex of the graph of the equation $y = -2x^2 + x - 5$?

(A) $-\frac{1}{4}$     (B) $-6\frac{1}{4}$     (C) $4\frac{1}{4}$

(D) $5\frac{1}{4}$     (E) $-4\frac{7}{8}$

**6.** *Multiple Choice*   Evaluate $y = -2x^2 + 11x + 9$ when $x = -2$.

(A) $y = 21$     (B) $y = 22$

(C) $y = -21$     (D) $y = 5$

(E) $y = -5$

**7.** *Multiple Choice*   Which of the following quadratic equations is represented by the graph?

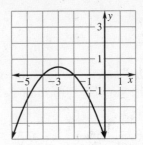

(A) $y = -2x^2 - 3x - 4$

(B) $y = \frac{1}{2}x^2 - 3x - 4$

(C) $y = 2x^2 - 3x - 4$

(D) $y = -\frac{1}{2}x^2 - 3x - 4$

(E) $y = \frac{1}{2}x^2 + 3x - 4$

***Quantitative Comparison***   In Exercises 8–11, find the coordinates of the vertex of the equation. Then choose the statement below that is true about the given values.

(A)   The value of *x* is greater.

(B)   The value of *y* is greater.

(C)   The values of *x* and *y* are equal.

(D)   The relationship cannot be determined from the given information.

**8.** $y = x^2 - 16$

**9.** $y = 3x^2 - 10x$

**10.** $y = x^2 - 2x + 2$

**11.** $y = 3x^2 + 2x + 1$

**Algebra 1, Concepts and Skills**
Standardized Test Practice Workbook

# Standardized Test Practice

**For use with pages 526–531**

**TEST TAKING STRATEGY**   **Learn as much as you can about a test ahead of time, such as the types of questions and the topics that the test will cover.**

**1.** *Multiple Choice*   Use the graph to determine the roots of the equation.

Ⓐ  3 and $-3$

Ⓑ  1 and 4

Ⓒ  $-\frac{1}{2}$ and 3

Ⓓ  $\frac{1}{2}$ and 3

Ⓔ  4 and $-3$

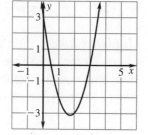

**2.** *Multiple Choice*   What are the $x$-intercepts of the graph of $y = x^2 - 3x - 18$?

Ⓐ  6 and $-3$   Ⓑ  $-6$ and 3

Ⓒ  9 and 2   Ⓓ  $-9$ and 2

Ⓔ  4 and 1

**3.** *Multiple Choice*   What are the $x$-intercepts of the graph of $y = 6x^2 - x - 1$?

Ⓐ  3 and 2   Ⓑ  $-3$ and 2

Ⓒ  $-\frac{1}{3}$ and $\frac{1}{2}$   Ⓓ  $-\frac{1}{3}$ and $-\frac{1}{2}$

Ⓔ  $\frac{1}{3}$ and $-\frac{1}{2}$

**4.** *Multiple Choice*   What are the solutions of $x^2 + x - 6 = 0$?

Ⓐ  3 and 2   Ⓑ  $-3$ and 2

Ⓒ  $\frac{1}{3}$ and 2   Ⓓ  $\frac{1}{3}$ and $\frac{1}{2}$

Ⓔ  $-3$ and 0

**5.** *Multiple Choice*   Which one of the following is a solution of $\frac{1}{5}x^2 = 5$?

Ⓐ  0   Ⓑ  5   Ⓒ  10

Ⓓ  $-20$   Ⓔ  25

*Quantitative Comparison*   In Exercises 6–9, solve for the required information. Then choose the statement below that is true about the given information.

Ⓐ  The value of column A is greater.

Ⓑ  The value of column B is greater.

Ⓒ  The two values are equal.

Ⓓ  The relationship cannot be determined from the given information.

| | Column A | Column B |
|---|---|---|
| **6.** | The $x$-coordinate of the vertex of $y = x^2 - 4x - 5$ | The $y$-coordinate of the vertex of $y = x^2 - 4x - 5$ |
| **7.** | The $x$-coordinate of the vertex of $\frac{1}{2}x^2 = 8$ | The sum of the roots of $\frac{1}{2}x^2 = 8$ |
| **8.** | The sum of the roots of $x^2 - x - 2 = 0$ | The sum of the roots of $x^2 + 2x = 15$ |
| **9.** | The $x$-coordinate of the vertex of $-3x^2 + 15x - 5 = 1$ | The sum of the roots of $2x^2 - 5x - 3 = 0$ |

**10.** *Multiple Choice*   Choose the equation of the parabola shown in the graph below.

Ⓐ  $y = 9x^2 - 1$

Ⓑ  $y = 9x^2 + 1$

Ⓒ  $y = \frac{1}{9}x^2 - 9$

Ⓓ  $y = 9x^2 + 1$

Ⓔ  $y = \frac{1}{9}x^2 - 1$

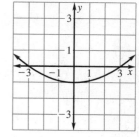

**11.** *Multiple Choice*   Which one of the following is a solution of $x^2 - x - 14 = 6$?

Ⓐ  0   Ⓑ  125   Ⓒ  $-3$

Ⓓ  $-4$   Ⓔ  4

# Standardized Test Practice

**For use with pages 533–539**

**TEST TAKING STRATEGY**   **Go back and check as much of your work as you can.**

**1.** *Multiple Choice*   Choose the correct form of the quadratic formula.

(A)   $x = \dfrac{b \pm \sqrt{b^2 + 4ac}}{2a}$

(B)   $x = \dfrac{-b \pm \sqrt{b^2 - 4ac}}{2a}$

(C)   $x = \dfrac{-a \pm \sqrt{a^2 - 4bc}}{2a}$

(D)   $x = \dfrac{-c \pm \sqrt{b^2 - 4ac}}{2b}$

(E)   $x = \dfrac{-b \pm \sqrt{a^2 - 4bc}}{2a}$

**2.** *Multiple Choice*   Choose the correct values of $a$, $b$, and $c$ in the equation $5x^2 - x + 6 = 0$.

(A)   $a = 5$, $b = 1$, $c = 6$

(B)   $a = -5$, $b = -1$, $c = -6$

(C)   $a = 5$, $b = 0$, $c = 6$

(D)   $a = 5$, $b = -1$, $c = 0$

(E)   $a = 5$, $b = -1$, $c = 6$

**3.** *Multiple Choice*   What are the $x$-intercepts of the graph of $y = 2x^2 - x - 15$?

(A)   $\frac{5}{2}, \frac{-5}{2}$     (B)   $-\frac{5}{2}, 3$     (C)   $\frac{5}{3}, -3$

(D)   $-5, 3$     (E)   $-5, \frac{15}{2}$

**4.** *Multiple Choice*   Which of the following is a solution of $2x^2 - 3x - 5 = 0$?

(A)   $1$     (B)   $-\frac{5}{2}$     (C)   $\frac{5}{2}$

(D)   $\frac{2}{5}$     (E)   $2$

**5.** *Multiple Choice*   Which of the following is a solution of $6x^2 - 7x - 5 = 0$?

(A)   $\frac{1}{2}$     (B)   $\frac{5}{3}$     (C)   $-2$

(D)   $-\frac{3}{5}$     (E)   $-1$

**6.** *Multiple Choice*   Which of the following is a solution of $7x^2 + 5x + 8 = 10$?

(A)   $\frac{2}{7}$     (B)   $1$     (C)   $\frac{7}{2}$

(D)   $-\frac{2}{7}$     (E)   $\frac{3}{7}$

**7.** *Multiple Choice*   You drop a rock off a bridge 30 feet above the ground into a stream. How long does it take the rock to hit the water? Round your answer to the nearest hundredth.

(A)   1.45 sec     (B)   1.88 sec

(C)   1.50 sec     (D)   2.10 sec

(E)   1.37 sec

**8.** *Multiple Choice*   An eagle circling a field at a height of 250 feet sees a rabbit below. The eagle dives at an initial speed of 110 feet per second. Estimate the time the rabbit has to escape. Round your answer to the nearest tenth.

(A)   1.7 sec     (B)   1.8 sec

(C)   1.6 sec     (D)   1.9 sec

(E)   2.0 sec

**9.** *Multi-Step Problem*   You are on a ski-lift 50 feet high. While the lift is stopped to let people off, you accidentally knock your keys out of your pocket. They fall off the seat towards the ground.

**a.** Write a vertical motion model for the path of the keys.

**b.** How long will it take for the keys to hit the ground?

**c.** *Critical Thinking*   What factors would change the path of the dropped keys?

# Standardized Test Practice

**For use with pages 540–545**

**TEST TAKING STRATEGY**  **Spend no more than a few minutes on each question.**

1. **Multiple Choice**  What is the discriminant of the equation $7x^2 - 3x + 10 = 0$?

    Ⓐ  271          Ⓑ  −271          Ⓒ  289

    Ⓓ  −289          Ⓔ  −277

2. **Multiple Choice**  What is the discriminant of the equation $-3x^2 - 12x + 17 = 8$?

    Ⓐ  36          Ⓑ  348          Ⓒ  −348

    Ⓓ  −252          Ⓔ  252

3. **Multiple Choice**  Use the discriminant to determine the number of solutions for the equation $3x^2 - 7x - 1 = 0$.

    Ⓐ  3          Ⓑ  1          Ⓒ  2

    Ⓓ  Infinitely many          Ⓔ  None

4. **Multiple Choice**  Use the discriminant to determine the number of solutions for the equation $\frac{1}{2}x^2 = 8$.

    Ⓐ  3          Ⓑ  1          Ⓒ  2

    Ⓓ  Infinitely many          Ⓔ  None

5. **Multiple Choice**  What effect does increasing the value of $c$ by 2 have on the number of solutions for the graph of $5x^2 - 2x - 1 = 0$?

    Ⓐ  Increases the number of solutions by 2

    Ⓑ  Decreases the number of solutions by 2

    Ⓒ  Increases the number of solutions by 1

    Ⓓ  Has no effect on the number of solutions

    Ⓔ  Decreases the number of solutions by 1.

6. **Multiple Choice**  What effect does decreasing the value of $c$ by 3 have on the number of solutions to the graph of $3x^2 + 6x + 3 = 0$?

    Ⓐ  Increases the number of solutions by 2

    Ⓑ  Decreases the number of solutions by 2

    Ⓒ  Increases the number of solutions by 1

    Ⓓ  Decreases the number of solutions by 1

    Ⓔ  Has no effect on the number of solutions

7. **Multiple Choice**  For the equation $2x^2 - 5x + 6 = 0$, the graph of the equation would __?__ .

    Ⓐ  Have one $x$-intercept

    Ⓑ  Have two $x$-intercepts

    Ⓒ  Have no $x$-intercepts

    Ⓓ  Have no $y$-intercepts

    Ⓔ  None of these

*Quantitative Comparison*  In Exercises 8–10, determine the number of solutions for each equation. Then choose the statement below that is true about the number of solutions.

   Ⓐ  The number of solutions in column A is greater.

   Ⓑ  The number of solutions in column B is greater.

   Ⓒ  The number of solutions is equal.

   Ⓓ  The relationship cannot be determined from the given information.

| | Column A | Column B |
|---|---|---|
| 8. | $3x^2 - 2x - 6 = 0$ | $-3x^2 + 2x - 6 = 0$ |
| 9. | $\frac{1}{2}x^2 - 18 = 0$ | $-5x^2 + 10 = 0$ |
| 10. | $x^2 - 2x + 1 = 0$ | $5x^2 - 8x + 3 = 0$ |

NAME _____ DATE _____

# *Standardized Test Practice*

**TEST TAKING STRATEGY**   **If you can, check your answer using a different method than you used originally to avoid making the same mistake twice.**

1. *Multiple Choice*   Which ordered pair is a solution of the inequality $y \leq 3x^2 - 5x - 6$?

   Ⓐ $(2, -3)$   Ⓑ $(1, -8)$   Ⓒ $(3, 7)$
   Ⓓ $(-2, 18)$   Ⓔ $(0, 0)$

2. *Multiple Choice*   Which ordered pair is *not* solution of the inequality $y > 8x^2 - 2x - 3$?

   Ⓐ $(-2, 35)$   Ⓑ $(0, -2)$   Ⓒ $(1, 8)$
   Ⓓ $(1, 5)$   Ⓔ $(-1, 6)$

3. *Multiple Choice*   Identify the graph of $y \geq 5x^2 + 14x - 3$.

   Ⓐ 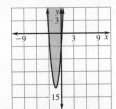   Ⓑ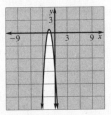

   Ⓒ 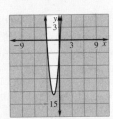   Ⓓ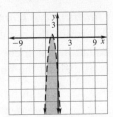

   Ⓔ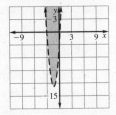

4. *Multiple Choice*   Choose the inequality which is represented by the graph.

   Ⓐ $y > 2x^2 - x - 4$
   Ⓑ $y \geq 2x^2 - x - 4$
   Ⓒ $y \leq 2x^2 - x - 4$
   Ⓓ $y < 2x^2 - x - 4$
   Ⓔ $y < -2x^2 - x - 4$

   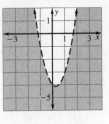

5. *Multiple Choice*   A volcano is catapulting rocks from its crater. The rocks traveling the farthest are following a path along $y = -x^2 + 60x - 80$, where $x$ and $y$ are measured in feet. To be safe from the falling rocks, you should be standing in the area defined by which of the inequalities?

   Ⓐ $y > -x^2 + 60x - 80$
   Ⓑ $y \geq -x^2 + 60x - 80$
   Ⓒ $y < -x^2 + 60x - 80$
   Ⓓ $y \leq -x^2 + 60x - 80$
   Ⓔ None of these

6. *Multi-Step Problem*   A backhoe is digging a trench. On its first sweep, the shovel follows a path modeled by the equation $y = x^2 - 2x - 3$ where $x$ is the length in feet and $y$ is the depth in feet of the hole.

   a. What is the deepest point of the hole?

   b. What is the length of the hole?

   c. *Critical Thinking*   A water line runs perpendicular to the ditch the backhoe is digging. If the line is buried $3\frac{1}{2}$ feet down, did the backhoe hit it on the first sweep of the shovel? Explain.

# Standardized Test Practice

**For use with pages 568–573**

**TEST TAKING STRATEGY**  **Read all of the answer choices before deciding which is the correct one.**

1. **Multiple Choice**  Classify the equation $7 + 5x + 2x^3 + 3x^2$ by degree and by the number of terms.

   Ⓐ  Quadratic polynomial

   Ⓑ  Quadratic binomial

   Ⓒ  Cubic binomial

   Ⓓ  Cubic trinomial

   Ⓔ  Cubic polynomial

2. **Multiple Choice**  Which of the following is equal to $(7x^2 + 3) + (5x^2 + 8)$?

   Ⓐ  $12x^2 + 11$  Ⓑ  $12x^2 + 5$  Ⓒ  $12x^2$

   Ⓓ  $12x^4 - 5$  Ⓔ  $12x^4 + 11$

3. **Multiple Choice**  Which of the following is equal to $(3x^2 + 7x - 6) + (5x^3 + 3x^2 + 10x - 8)$?

   Ⓐ  $8x^3 + 10x^2 + 10x - 14$

   Ⓑ  $8x^3 + 10x^2 + 10x + 2$

   Ⓒ  $8x^3 + 3x^2 + 17x - 14$

   Ⓓ  $5x^3 + 6x^2 + 17x - 14$

   Ⓔ  $5x^3 + 6x^2 + 17x + 2$

4. **Multiple Choice**  Which of the following is equal to $(4x^2 + 7x - 6) - (3x^2 + 9x - 8)$?

   Ⓐ  $x^2 + 2x + 2$    Ⓑ  $x^2 - 2x + 2$

   Ⓒ  $x^2 - 2x - 14$    Ⓓ  $7x^2 + 18x - 14$

   Ⓔ  $7x^2 + 18x + 2$

5. **Multiple Choice**  Which of the following is equal to $(5x^3 + 2x^2 - 7x + 9) - (-4x^3 + 3x^2 + 2x + 1)$?

   Ⓐ  $x^3 - x^2 - 5x - 8$

   Ⓑ  $9x^3 - x^2 - 9x + 8$

   Ⓒ  $x^3 + 5x^2 - 9x + 1$

   Ⓓ  $9x^3 + 5x^2 - 9x + 1$

   Ⓔ  $x^3 - x^2 - 9x + 8$

6. **Multiple Choice**  You want to fence in a pool and deck area. Find the expression which correctly determines the amount of fencing required to complete the job.

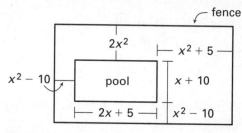

   Ⓐ  $5x^2 + 3x$

   Ⓑ  $10x^2 + 6x$

   Ⓒ  $10x^2 + 3x + 10$

   Ⓓ  $5x^2 + 6x + 10$

   Ⓔ  $10x^2 + 8x + 20$

7. **Multi-Step Problem**  A company is keeping track of 3 different divisions. The sales $S$ in thousands of dollars for the first two divisions and the whole company are modeled below, where $t$ is the number of years since 1998.

   Division 1:  $S = 12.2t^2 + 6.3t + 65$

   Division 2:  $S = 13.5t^2 - 4.3t + 75$

   Company:  $S = 34.6t^2 + 3.2t + 213$

   **a.**  Find a model that represents division 3.

   **b.**  Predict sales for division 3 in 2000.

   **c.**  *Critical Thinking*  Sales are $350,000 one year, but division 2 did not contribute. Write an equation in terms of $t$ to model this situation.

NAME _____    DATE _____

# *Standardized Test Practice*

For use with pages 575–580

TEST TAKING STRATEGY    **Before you give up on a question, try to eliminate some of your choices so you can make an educated guess.**

1. *Multiple Choice*  Which of the following is equal to $3x^2(5x - x^2 + 7)$?

   (A)  $8x^3 - 3x^4 + 21x^2$

   (B)  $-3x^4 + 8x^2 + 21x^2$

   (C)  $-3x^4 + 8x^3 + 10x^2$

   (D)  $-3x^4 + 15x^3 + 21x^2$

   (E)  $3x^4 + 15x^3 + 21x^2$

2. *Multiple Choice*  Which of the following is equal to $-2y^3(2y^2 - 5y + 6)$?

   (A)  $4y^2 - 10y^4 + 12y^3$

   (B)  $-4y^6 + 10y^4 - 12y^3$

   (C)  $-4y^6 - 10y^4 - 6y^3$

   (D)  $y^5 - 7y^4 + 4y^3$

   (E)  $-4y^5 + 10y^4 - 12y^3$

3. *Multiple Choice*  Which of the following is equal to $(b - 8)(b + 2)$?

   (A)  $b^2 - 6b - 16$   (B)  $b^2 + 6b - 16$

   (C)  $b^2 - 6b - 6$   (D)  $b^2 - 6b + 6$

   (E)  $b^2 + 10b - 16$

4. *Multiple Choice*  Which of the following is equal to $(x - 14)(x - 3)$?

   (A)  $x^2 - 17 - 17$   (B)  $x^2 - 11x + 42$

   (C)  $x^2 - 17x - 42$   (D)  $x^2 - 11x - 42$

   (E)  $x^2 - 17x + 42$

5. *Multiple Choice*  Find the product of $(x + 5)(3x^2 - 2x + 1)$.

   (A)  $3x^3 + 13x^2 - 10x + 5$

   (B)  $3x^3 + 13x^2 - 10x + 6$

   (C)  $3x^3 + 6x^2 - 6x + 6$

   (D)  $3x^3 + 13x^2 - 9x + 5$

   (E)  $3x^3 + 17x^2 - 9x + 5$

6. *Multiple Choice*  Find the product of $(4x^2 + 6x - 7)(2x - 3)$.

   (A)  $8x^3 - 24x^2 - 32x + 21$

   (B)  $8x^3 - 12x^2 - 20x + 21$

   (C)  $8x^3 - 32x + 21$

   (D)  $8x^3 - 4x - 21$

   (E)  $8x^3 + 24x^2 - 32x + 21$

7. *Multiple Choice*  Find the area of the parallelogram by using the formula  $A = bh$.

   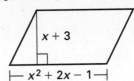

   (A)  $x^3 + 5x^2 + 5x - 3$

   (B)  $x^3 + 6x^2 - 6x + 3$

   (C)  $x^3 + 2x^2 - x + 3$

   (D)  $x^2 + 3x + 2$

   (E)  $x^2 + 3x - 2$

*Quantitative Comparison*   In Exercises 8–10, evaluate the expression for $x = 2$ and $y = -3$. Then choose the statement below that is true about the results.

   (A)  The number in column A is greater.

   (B)  The number in column B is greater.

   (C)  The two numbers are equal.

   (D)  The relationship cannot be determined from the given information.

| | Column A | Column B |
|---|---|---|
| 8. | $2x(x^2 + 3x + 1)$ | $-2y(3y^2 + 2y)$ |
| 9. | $(x + 1)(x - 6)$ | $(3y + 2)(2y + 5)$ |
| 10. | $(2x - 9)(x - 5)$ | $(y + 3)(y^2 + 2y + 5)$ |

**Algebra 1, Concepts and Skills**
Standardized Test Practice Workbook

NAME _____ DATE _____

# Standardized Test Practice

For use with pages 581–587

**TEST TAKING STRATEGY** **Go back and check as much of your work as you can.**

1. **Multiple Choice** Which of the following is equal to $(x + 3)^2$?

   Ⓐ $x^2 + 9$      Ⓑ $x^2 - 6x + 9$

   Ⓒ $x^2 - 9$      Ⓓ $x^2 - 6x - 9$

   Ⓔ $x^2 + 6x + 9$

2. **Multiple Choice** Which of the following is equal to $(2x - 3)^2$?

   Ⓐ $4x^2 + 9$      Ⓑ $4x^2 - 12x + 9$

   Ⓒ $4x^2 - 9$      Ⓓ $4x^2 - 12x - 9$

   Ⓔ $4x^2 - 10x + 9$

3. **Multiple Choice** Which of the following is equal to the expression $25x^2 - 70x + 49$?

   Ⓐ $(7x + 5)^2$      Ⓑ $(5x + 7)^2$

   Ⓒ $(7x - 5)^2$      Ⓓ $(5x - 7)^2$

   Ⓔ $(5x - 7)(5x + 7)$

4. **Multiple Choice** Find the area of the circle below. (*Hint:* $A = \pi r^2$.)

   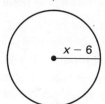

   $x - 6$

   Ⓐ $A = \pi(x^2 + 6)$

   Ⓑ $A = \pi(x^2 + 12x + 12)$

   Ⓒ $A = \pi(x^2 - 12x - 12)$

   Ⓓ $A = \pi(x^2 + 12x + 36)$

   Ⓔ $A = \pi(x^2 - 12x + 36)$

5. **Multiple Choice** Which of the following is equal to $(5x + 3)(5x - 3) - (2x - 5)^2$?

   Ⓐ $21x^2 - 20x + 16$  Ⓑ $29x^2 - 20x + 16$

   Ⓒ $21x^2 - 20x - 34$  Ⓓ $21x^2 - 34$

   Ⓔ $21x^2 + 20x - 34$

6. **Multiple Choice** Which of the following represents the area of the figure shown?

   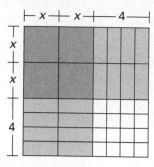

   Ⓐ $4x^2 - 16$

   Ⓑ $4x^2 + 16x + 16$

   Ⓒ $4x^2 - 16x + 16$

   Ⓓ $4x^2 + 8x - 16$

   Ⓔ $4x + 8$

*Quantitative Comparison* In Exercises 7–9, evaluate the expression for the given values. Then choose the statement below that is true about the results.

   Ⓐ The number in column A is greater.

   Ⓑ The number in column B is greater.

   Ⓒ The two numbers are equal.

   Ⓓ The relationship cannot be determined from the given information.

| | Column A | Column B |
|---|---|---|
| 7. | $(x + 2)^2$ when $x = 3$ | $(y - 2)^2$ when $y = 3$ |
| 8. | $(a - b)^2$ when $a = 8$ and $b = -4$ | $(a + b)^2$ when $a = 8$ and $b = -4$ |
| 9. | $(a - b)(a + b)$ when $a = 5$ and $b = -4$ | $(a^2 - b^2)^2$ when $a = 5$ and $b = -4$ |

# *Standardized Test Practice*

For use with pages 588–593

**TEST TAKING STRATEGY**   **Some questions involve more than one step. Reading too quickly might lead to mistaking the answer to a preliminary step for your final answer.**

1. *Multiple Choice*   Which of the following is one of the solutions of the equation $(x - 3)(x + 8) = 0$?

   Ⓐ 8          Ⓑ $-3$          Ⓒ 0

   Ⓓ $\frac{3}{8}$          Ⓔ 3

2. *Multiple Choice*   Choose the solutions of the equation $(4x - 3)(x + 1) = 0$.

   Ⓐ $1, -\frac{3}{4}$          Ⓑ $-1, \frac{3}{4}$          Ⓒ $1, -\frac{4}{3}$

   Ⓓ $1, \frac{4}{3}$          Ⓔ $-1, \frac{4}{3}$

3. *Multiple Choice*   Which of the following are solutions of the equation $(2x - 5)^2 = 0$?

   Ⓐ $\frac{5}{2}$          Ⓑ $\frac{5}{2}, -\frac{5}{2}$          Ⓒ $\frac{2}{5}$

   Ⓓ $\frac{2}{5}, -\frac{2}{5}$          Ⓔ 5

4. *Multiple Choice*   Choose the solutions of the equation $(3x - 1)(x + 6)(2x + 7) = 0$.

   Ⓐ $\frac{1}{3}, 6, \frac{7}{2}$          Ⓑ $3, -6, -7$

   Ⓒ $\frac{1}{3}, -6, -\frac{7}{2}$  Ⓓ $-\frac{1}{3}, 6, \frac{7}{2}$

   Ⓔ $3, -6, -\frac{2}{7}$

5. *Multiple Choice*   Choose the equation whose graph is shown.

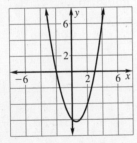

   Ⓐ $y = (x + 2)(x + 3)$

   Ⓑ $y = (x - 2)(x + 3)$

   Ⓒ $y = (x - 3)(x + 2)$

   Ⓓ $y = (x - 3)^2$

   Ⓔ $y = (x - 3)(x - 2)$

6. *Multiple Choice*   What are the coordinates of the vertex of the graph of $y = (x - 4)(2x + 1)$?

   Ⓐ $\left(-4, \frac{1}{2}\right)$          Ⓑ $\left(\frac{5}{3}, -8\frac{1}{3}\right)$

   Ⓒ $\left(-\frac{4}{3}, -8\frac{1}{3}\right)$          Ⓓ $\left(\frac{7}{4}, -10\frac{1}{8}\right)$

   Ⓔ $\left(4, -\frac{1}{2}\right)$

7. *Multiple Choice*   What are the coordinates of the vertex of the graph of $y = (-2x + 1)(x - 3)$?

   Ⓐ $(0, -3)$          Ⓑ $\left(\frac{3}{2}, 3\right)$

   Ⓒ $\left(\frac{7}{4}, 3\frac{1}{8}\right)$          Ⓓ $(3, 0)$

   Ⓔ $(2, 3)$

8. *Multi-Step Problem*   A dome tent's arch is modeled by $y = -0.18(x - 6)(x + 6)$, where $x$ and $y$ are measured in feet.

   a. How wide is the tent at the base?

   b. How high is the tent at its highest point?

   c. *Critical Thinking*   The base of the tent is a square. Each person requires 18 square feet to sleep comfortably. If you require an area of 25 square feet to store gear and supplies, how many people can sleep in the tent?

NAME _____ DATE _____

# *Standardized Test Practice*

For use with pages 595–601

**TEST TAKING STRATEGY** **Be aware of how much time you have left, but keep focused on your work.**

1. *Multiple Choice* Which of the following is a correct factorization of $x^2 + 4x - 12$?

  Ⓐ $(x + 3)(x - 4)$  Ⓑ $(x + 6)(x - 2)$

  Ⓒ $(x - 6)(x + 2)$  Ⓓ $(x - 3)(x + 4)$

  Ⓔ Cannot be factored

2. *Multiple Choice* Which of the following is a correct factorization of $x^2 - 5x + 6$?

  Ⓐ $(x - 6)(x - 1)$  Ⓑ $(x + 6)(x - 1)$

  Ⓒ $(x - 2)(x - 3)$  Ⓓ $(x + 2)(x + 3)$

  Ⓔ Cannot be factored

3. *Multiple Choice* Which of the following is a correct factorization of $x^2 + 10x - 9$?

  Ⓐ $(x - 9)(x + 1)$  Ⓑ $(x + 9)(x - 1)$

  Ⓒ $(x - 9)(x - 1)$  Ⓓ $(x - 3)(x - 3)$

  Ⓔ Cannot be factored

4. *Multiple Choice* Use factoring to solve the equation $x^2 + 4x = 21$.

  Ⓐ $-7, -3$  Ⓑ $7, 3$  Ⓒ $-7, 3$

  Ⓓ $7, -3$  Ⓔ Cannot be factored

5. *Multiple Choice* Use factoring to solve the equation $x^2 + 2x - 20 = 28$.

  Ⓐ $6, -8$  Ⓑ $6, 8$

  Ⓒ $4, -12$  Ⓓ $-4, 12$

  Ⓔ Cannot be factored

6. *Multiple Choice* Use factoring to solve the equation $x^2 + 13x + 12 = -24$.

  Ⓐ $9, -4$  Ⓑ $9, 4$

  Ⓒ $-9, 4$  Ⓓ $-9, -4$

  Ⓔ Cannot be factored

7. *Multiple Choice* The area of a rectangle is given by $x^2 - 4x - 21$. Use factoring to find expressions for the dimensions of the rectangle.

  Ⓐ $x - 7, x - 3$  Ⓑ $x + 7, x - 3$

  Ⓒ $x - 21, x + 1$  Ⓓ $x - 7, x + 3$

  Ⓔ Cannot be determined

8. *Multiple Choice* The area of a rectangle is 48. If the length is 8 less than its width, what is its width?

  Ⓐ 6  Ⓑ 4  Ⓒ 12

  Ⓓ 8  Ⓔ 9

*Quantitative Comparison* In Exercises 9–12, solve the equation by factoring and add the solutions. Then choose the statement that is true about the values.

  Ⓐ The value in column A is greater.

  Ⓑ The value in column B is greater.

  Ⓒ The two values are equal.

  Ⓓ The relationship cannot be determined from the given information.

| | Column A | Column B |
|---|---|---|
| 9. | $x^2 - 3x - 14 = 14$ | $x^2 + 5x + 6 = 0$ |
| 10. | $x^2 - 2x - 20 = 28$ | $x^2 + 5x - 36 = 0$ |
| 11. | $x^2 - 4x + 10 = 15$ | $x^2 - 4x - 12 = 0$ |
| 12. | $x^2 - 4x - 12 = 0$ | $x^2 = 25$ |

NAME _____ DATE _____

## *Standardized Test Practice*

For use with pages 603–608

**TEST TAKING STRATEGY** **As soon as the test begins, start working. Keep a steady pace and stay focused on the test.**

1. *Multiple Choice* Which of the following is a correct factorization of $10x^2 + 19x + 6$?

  Ⓐ $(2x + 2)(5x + 3)$

  Ⓑ $(5x + 2)(2x + 3)$

  Ⓒ $(5x - 3)(2x + 2)$

  Ⓓ $(10x + 6)(x + 1)$

  Ⓔ $(10x + 1)(x + 6)$

2. *Multiple Choice* Which of the following is a correct factorization of $21x^2 + 8x - 4$?

  Ⓐ $(21x - 4)(x + 1)$

  Ⓑ $(7x - 2)(3x - 2)$

  Ⓒ $(7x + 2)(3x - 2)$

  Ⓓ $(7x - 2)(3x + 2)$

  Ⓔ $(7x - 4)(3x + 1)$

3. *Multiple Choice* Which of the following is a correct factorization of $27x^2 - 69x + 40$?

  Ⓐ $(3x - 5)(9x + 8)$

  Ⓑ $(9x - 5)(3x - 8)$

  Ⓒ $(3x - 5)(9x - 8)$

  Ⓓ $(3x - 4)(9x - 10)$

  Ⓔ $(9x - 4)(3x - 10)$

4. *Multiple Choice* Which of the following is a solution of the equation $14x^2 - 23x - 30 = 0$?

  Ⓐ $-\frac{5}{2}$      Ⓑ $\frac{5}{2}$      Ⓒ $\frac{6}{7}$

  Ⓓ $-\frac{2}{7}$      Ⓔ $\frac{15}{2}$

5. *Multiple Choice* Which of the following is a solution of the equation $24x^2 + 26x = 5$?

  Ⓐ $\frac{1}{6}$      Ⓑ $\frac{4}{5}$      Ⓒ $\frac{5}{4}$

  Ⓓ $\frac{5}{6}$      Ⓔ $-\frac{1}{4}$

6. *Multiple Choice* Which one of the following equations cannot be solved by factoring with integer coefficients?

  Ⓐ $6x^2 + 5x - 6 = 0$

  Ⓑ $16x^2 - 26x = -3$

  Ⓒ $10x^2 + 9x + 3 = 0$

  Ⓓ $15x^2 = 12 - 8x$

  Ⓔ $6x^2 + 14 = 31x$

7. *Multiple Choice* Solve $33x^2 + 71x - 14 = 0$ by factoring.

  Ⓐ $\frac{7}{11}, \frac{2}{3}$      Ⓑ $\frac{7}{11}, -\frac{2}{3}$

  Ⓒ $12\frac{7}{11}, \frac{2}{3}$      Ⓓ $-\frac{2}{11}, \frac{7}{3}$

  Ⓔ $\frac{2}{11}, -\frac{7}{3}$

8. *Multiple Choice* The area of a square is given by $A = 4x^2 + 28x + 49$. The length of each side is ___?___ .

  Ⓐ $2x - 7$      Ⓑ $7x - 2$

  Ⓒ $7x + 2$      Ⓓ $2x + 7$

  Ⓔ $-2x + 7$

9. *Multi-Step Problem* A diver jumps off a 20 foot high diving board with an initial upward velocity of 4 feet per second. Use the vertical motion model $h = -16t^2 + vt + s$.

  a. Use the information above to model the vertical motion of the diver when the diver enters the water.

  b. Find the solutions by factoring the model you found in part (a).

  c. *Critical Thinking* Explain why only one of the solutions is reasonable.

# Standardized Test Practice

For use with pages 609–615

**TEST TAKING STRATEGY**   Think positively during a test. This will help keep up your confidence and enable you to focus on each question.

1. *Multiple Choice*   Which of the following is a factorization of $x^2 - 16x + 64$?

   Ⓐ  $(x + 8)(x - 8)$   Ⓑ  $(x + 8)^2$

   Ⓒ  $2(x - 4)(x - 4)$   Ⓓ  $(x - 8)^2$

   Ⓔ  $2(x + 4)(x - 4)$

2. *Multiple Choice*   Which of the following is a factorization of $25x^2 - 16$?

   Ⓐ  $(5x - 4)(5x + 4)$   Ⓑ  $(5x - 4)^2$

   Ⓒ  $(5x - 16)(5x - 1)$   Ⓓ  $(5x + 4)^2$

   Ⓔ  $(5x + 16)(5x + 1)$

3. *Multiple Choice*   Which of the following is a factorization of $36x^2 + 84x + 49$?

   Ⓐ  $(6x - 7)(6x + 7)$   Ⓑ  $(6x + 7)^2$

   Ⓒ  $2(9x + 3)(2x + 4)$   Ⓓ  $(6x - 7)^2$

   Ⓔ  Cannot be factored

4. *Multiple Choice*   Which of the following is a factorization of $20x^2 - 125$?

   Ⓐ  $5(2x - 10)(2x + 10)$

   Ⓑ  $5(2x - 5)^2$

   Ⓒ  $5(2x - 5)(2x + 5)$

   Ⓓ  $5(2x + 5)^2$

   Ⓔ  Cannot be factored

5. *Multiple Choice*   Which of the following is a factorization of $-75x^2 - 30x - 3$?

   Ⓐ  $-3(5x + 1)(5x - 1)$   Ⓑ  $-3(5x - 1)^2$

   Ⓒ  $(15x - 3)(-5x - 1)$

   Ⓓ  $-3(5x + 1)^2$

   Ⓔ  Cannot be factored

6. *Multiple Choice*   Which of the following is a solution of $2x^2 - 98 = 0$?

   Ⓐ  0   Ⓑ  14   Ⓒ  12

   Ⓓ  7   Ⓔ  Cannot be factored

7. *Multiple Choice*   Which of the following is a solution of $4x^2 + \frac{4}{3}x + \frac{1}{9} = 0$?

   Ⓐ  $-\frac{1}{6}$   Ⓑ  $\frac{1}{6}$   Ⓒ  $-\frac{2}{3}$

   Ⓓ  $\frac{2}{3}$   Ⓔ  $\frac{3}{2}$

8. *Multiple Choice*   An object is propelled from the ground with an initial upward velocity of 80 feet per second. How long does it take to reach a height of 100 feet?

   Ⓐ  1.9 sec   Ⓑ  2.6 sec

   Ⓒ  1.8 sec   Ⓓ  2.1 sec

   Ⓔ  2.5 sec

9. *Multiple Choice*   Determine the diameter $D$ necessary for a wire rope to lift a 10.24 ton load safely. Use the model $4 \cdot D^2 = S$ where $D$ is the diameter of the rope in inches and $S$ is the safe working load in tons.

   Ⓐ  4 inches   Ⓑ  1.6 inches

   Ⓒ  1.8 inches   Ⓓ  2 inches

   Ⓔ  6.4 inches

*Quantitative Comparison*   In Exercises 10 and 11, solve the equation by factoring and add the solutions. Then choose the statement that is true about the values.

   Ⓐ  The value in column A is greater.

   Ⓑ  The value in column B is greater.

   Ⓒ  The two values are equal.

   Ⓓ  The relationship cannot be determined from the given information.

|     | *Column A* | *Column B* |
|-----|------------|------------|
| 10. | $x^2 - 16 = 0$ | $9x^2 - 25 = 0$ |
| 11. | $36x^2 - 16 = 0$ | $x^2 - 10x + 25 = 0$ |

# LESSON
## 10.8

# *Standardized Test Practice*

**For use with pages 616–622**

**TEST TAKING STRATEGY**   **Go back and check as much of your work as you can.**

1. *Multiple Choice*   What is the greatest common factor of $3x^2yz + 12xz$ ?

    **(A)** $xz$      **(B)** $3x^2z$      **(C)** $3x^2$

    **(D)** $3x$      **(E)** $3xz$

2. *Multiple Choice*   Find the greatest common factor and factor it out of $4a^3 - 2a^2 + 8$ .

    **(A)** $2(2a^3 - 2a^2 + 4)$

    **(B)** $2(2a^3 - a^2 + 4)$

    **(C)** $2a^2(2a - 1 + 4)$

    **(D)** $a^2(4a - 1 + 4)$

    **(E)** None of these

3. *Multiple Choice*   Factor $x^3 + x^2 + x + 1$ by grouping.

    **(A)** $(x + 1)(x^2 + 1)$

    **(B)** $(x + 1)(x + 1)$

    **(C)** $(x + 1)^2$

    **(D)** $x^2(x + 1)$

    **(E)** $x^2(x + 1)(x + 1)^2$

4. *Multiple Choice*   Factor $x^3 - 27$ .

    **(A)** $(x - 3)(x^2 + 3x + 3)$

    **(B)** $(x - 3)(x^2 + 3x + 9)$

    **(C)** $(x - 3)(x^2 - 3x + 9)$

    **(D)** $(x - 3)(x^2 + 3x - 9)$

    **(E)** $(x + 3)(x^2 - 3x + 9)$

5. *Multiple Choice*   Factor $2x^2 - xy + 4x - 2y$ .

    **(A)** $(2x - y)(x - 2)$

    **(B)** $(2x + y)(x - 2)$

    **(C)** $(2x - y)(x + 2)$

    **(D)** $(2x - y)(2x - 1)$

    **(E)** None of these

6. *Multiple Choice*   Solve $8x - 2x^3 = 0$ .

    **(A)** $0, 2, 4$      **(B)** $0, 2$

    **(C)** $-2, 2$      **(D)** $-2, 0, 2$

    **(E)** None of these

7. *Multi-Step Problem*   An object is dropped from a height of 128 feet. Its height is given by $h = -16t^2 + 128$ .

    **a.** How long does it take to reach halfway to the ground?

    **b.** How long does it take to reach the ground?

**Algebra 1, Concepts and Skills**
Standardized Test Practice Workbook

**TEST TAKING STRATEGY**   Some questions involve more than one step. Reading too quickly might lead to mistaking the answer to a preliminary step for your final answer.

1. **Multiple Choice**   What are the means of the proportion $\frac{5}{8} = \frac{10}{16}$?

   (A) 5 and 8        (B) 5 and 16

   (C) 5 and 10       (D) 8 and 16

   (E) 8 and 10

2. **Multiple Choice**   Which of the following is the solution of $\frac{5}{16} = \frac{x}{12}$?

   (A) $6\frac{2}{3}$      (B) $\frac{4}{15}$      (C) $3\frac{3}{4}$

   (D) $3\frac{1}{2}$      (E) $1\frac{1}{16}$

3. **Multiple Choice**   Which of the following is the solution of $\frac{4}{3x} = \frac{14}{21}$?

   (A) 2      (B) $\frac{1}{2}$      (C) 8

   (D) $\frac{1}{8}$      (E) 9

4. **Multiple Choice**   Which of the following is a solution of $\frac{2x}{5} = \frac{40}{x}$?

   (A) 25      (B) 50      (C) 4.5

   (D) 10      (E) 20

5. **Multiple Choice**   Which of the following is solution of $\frac{x-2}{10} = \frac{x}{15}$?

   (A) $\frac{2}{5}$      (B) 6      (C) −6

   (D) $-\frac{2}{5}$      (E) 7

6. **Multiple Choice**   Which of the following is a solution of $\frac{-3}{y-3} = \frac{y}{-6}$?

   (A) 3      (B) −6      (C) 6

   (D) 2      (E) 4

7. **Multiple Choice**   You are looking at a map to see how much farther it is to your destination. The distance on the map is 1.6 inches. If the map has a scale of 1 mile to $\frac{1}{32}$ of an inch, how much farther do you have to go?

   (A) 51.2 miles      (B) 15.6 miles

   (C) 23.4 miles      (D) 20.0 miles

   (E) 57.5 miles

**Quantitative Comparison**   In Exercises 8–10, solve the proportion. Then choose the statement below that is true.

   (A)   The value in column A is greater.

   (B)   The value in column B is greater.

   (C)   The two values are equal.

   (D)   The relationship cannot be determined from the given information.

| | Column A | Column B |
|---|---|---|
| 8. | $\frac{3}{x} = \frac{10}{12}$ | $\frac{2x+1}{14} = \frac{7}{9}$ |
| 9. | $\frac{x+7}{15} = \frac{x}{6}$ | $\frac{x+2}{3} = \frac{1}{x}$ |
| 10. | $\frac{x+2}{10} = \frac{x}{x+3}$ | $\frac{x-3}{x} = \frac{6}{2x+5}$ |

NAME _____ DATE _____

# Standardized Test Practice

**For use with pages 639–644**

**TEST TAKING STRATEGY**   **Think positively during a test. This will help keep up your confidence and enable you to focus on each question.**

1. **Multiple Choice**   Which of the following equations models inverse variation?

   Ⓐ $y = \frac{1}{8}x$   Ⓑ $y = 8x$

   Ⓒ $y = x + 8$   Ⓓ $y = 8 - x$

   Ⓔ $xy = 8$

2. **Multiple Choice**   Which of the following equations models direct variation when $x = 7$ and $y = 21$?

   Ⓐ $y = 3x$   Ⓑ $y = -3x$

   Ⓒ $y = \frac{147}{x}$   Ⓓ $y = \frac{1}{3}x$

   Ⓔ $y = \frac{x}{147}$

3. **Multiple Choice**   The variables $x$ and $y$ vary inversely. When $x$ is 15, $y$ is 40. If $x$ is 5, then $y$ is __?__ .

   Ⓐ 13.3   Ⓑ 115   Ⓒ 20

   Ⓓ 120   Ⓔ 3000

4. **Multiple Choice**   The area of a rectangle is 15 square inches. Which equation shows the correct relationship of length to width?

   Ⓐ $y = \frac{x}{15}$   Ⓑ $y = 15x$

   Ⓒ $y = 15 - x$   Ⓓ $y = \frac{15}{x}$

   Ⓔ $y = 15 + x$

5. **Multiple Choice**   Which of the following would *not* be an example of inverse variation?

   Ⓐ The hours $h$ you must work to earn $500 and your hourly pay $p$.

   Ⓑ The base $b$ and height $h$ of a triangle if the area is 15 square inches.

   Ⓒ The time $t$ it takes to drive 30 miles and your rate $r$.

   Ⓓ The rectangle's length $l$ and the area $A$ if the width is 5 inches.

   Ⓔ None of these

6. **Multiple Choice**   Choose the type of variation and the equation that represents the graph.

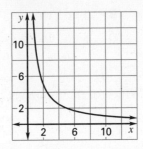

   Ⓐ Inverse; $y = 10x$

   Ⓑ Inverse; $y = \frac{10}{x}$

   Ⓒ Direct; $y = 10x$

   Ⓓ Direct; $y = \frac{10}{x}$

   Ⓔ None of these

7. **Multiple Choice**   The variables $x$ and $y$ vary inversely. When $x$ is $-3$, $y$ is 18. If $y$ is 6, then $x$ is __?__ .

   Ⓐ $-1$   Ⓑ 1   Ⓒ $-9$

   Ⓓ 9   Ⓔ 3

8. **Multi-Step Problem**   You would like to buy a new mountain bike that costs $1200.

   **a.** Make a table of possible values of the hours worked $h$ and pay rate $p$ to earn $1200. Sketch a graph.

   **b.** Does the model represent direct or inverse variation? Explain.

   **c.** *Critical Thinking*   Your friend is looking at a bike that costs $800. How does the model for this situation relate to the model for your situation?

**Algebra 1, Concepts and Skills**
Standardized Test Practice Workbook

# *Standardized Test Practice*

**For use with pages 646–651**

**TEST TAKING STRATEGY**    Before you give up on a question, try to eliminate some of your choices so you can make an educated guess.

**1.** *Multiple Choice*   What is the simplified form of the expression $\dfrac{3x^2 + 6x}{x^2 + 9x + 14}$?

  Ⓐ $\dfrac{3}{x+7}$    Ⓑ $\dfrac{x+2}{x+7}$    Ⓒ $\dfrac{3}{x+2}$

  Ⓓ $\dfrac{3x}{x+7}$    Ⓔ $\dfrac{x+3}{(x+7)(x+2)}$

**2.** *Multiple Choice*   What is the simplified form of the expression $\dfrac{6x^2 + 3x - 9}{6x^2 + 45x + 54}$?

  Ⓐ $\dfrac{x-3}{15x+18}$    Ⓑ $\dfrac{x-1}{x+6}$    Ⓒ $\dfrac{3(x-1)}{x+6}$

  Ⓓ $\dfrac{3(2x+3)}{x+6}$    Ⓔ $\dfrac{x-1}{(x+6)(2x+3)}$

**3.** *Multiple Choice*   What is the simplified form of the expression $\dfrac{2x^3 + 8x}{4x^3 - 16x}$?

  Ⓐ $\dfrac{x}{x+2}$    Ⓑ $\dfrac{1}{2}$    Ⓒ $\dfrac{1}{2x^3 - 2}$

  Ⓓ $\dfrac{x}{2}$    Ⓔ $\dfrac{x^2 + 4}{2(x+2)(x-2)}$

**4.** *Multiple Choice*   Simplify $\dfrac{4-x}{x^2 - 4x}$.

  Ⓐ $\dfrac{1}{x-1}$    Ⓑ $-\dfrac{1}{x-1}$

  Ⓒ $\dfrac{1}{x}$    Ⓓ $-\dfrac{1}{x}$

  Ⓔ Already in simplest form.

**5.** *Multiple Choice*   Simplify $\dfrac{x^2 - 4}{2x + 2}$.

  Ⓐ $x - 2$    Ⓑ $\dfrac{x-2}{2}$

  Ⓒ $x + 2$    Ⓓ $\dfrac{x+2}{2}$

  Ⓔ Already in simplest form.

**6.** *Multiple Choice*   What is the simplified form of the expression $\dfrac{-x^2 - 8x - 12}{x^2 + 2x - 24}$?

  Ⓐ $\dfrac{-4x-1}{2}$    Ⓑ $\dfrac{x+2}{x-4}$    Ⓒ $\dfrac{-x-2}{x-4}$

  Ⓓ $\dfrac{-1}{x-2}$    Ⓔ $\dfrac{x-6}{(x+6)(x-4)}$

**7.** *Multiple Choice*   Which ratio represents the ratio of the area of the smaller rectangle to the area of the larger rectangle?

  Ⓐ $\dfrac{x}{x+1}$      Ⓑ $\dfrac{1}{3(x+1)}$

  Ⓒ $\dfrac{x^2}{x^2 + 4x + 3}$      Ⓓ $\dfrac{x+3}{3x(x+1)}$

  Ⓔ $\dfrac{x}{(x+3)(x+1)}$

*Quantitative Comparison*   In Exercises 8–10, simplify the expression if possible and evaluate it for $x = 3$ and $y = -2$. Then choose the statement below that is true about the given values.

  Ⓐ   The value in column A is greater.

  Ⓑ   The value in column B is greater.

  Ⓒ   The two values are equal.

  Ⓓ   The relationship cannot be determined from the given information.

| | *Column A* | *Column B* |
|---|---|---|
| **8.** | $\dfrac{5x}{x^2 + 6x}$ | $\dfrac{7y^2}{14y - 7y^2}$ |
| **9.** | $\dfrac{x^2 - 4}{x^2 + 4x + 4}$ | $\dfrac{y^2 + 9}{y^2 + 6y + 9}$ |
| **10.** | $\dfrac{2x - 6}{x^2 - x - 6}$ | $\dfrac{y^3 + 2y^2}{y^4 - 3y^3 - 2y^2}$ |

**Algebra 1, Concepts and Skills**
Standardized Test Practice Workbook

*Chapter 11*

NAME _____ DATE _____

# Standardized Test Practice

**For use with pages 652–657**

**TEST TAKING STRATEGY**  **Go back and check as much of your work as you can.**

**1. Multiple Choice**  What is the simplified form of the expression $\dfrac{5x^3}{7x^2} \cdot \dfrac{21x^2}{20x}$?

(A) $\dfrac{3x^3}{4}$  (B) $\dfrac{x^2}{2}$  (C) $\dfrac{x^3}{4}$

(D) $\dfrac{3x^3}{4x}$  (E) $\dfrac{3x^2}{4}$

**2. Multiple Choice**  Which product equals the quotient $(5x - 1) \div \dfrac{x^2 + 4}{x - 1}$?

(A) $\dfrac{1}{(5x - 1)} \cdot \dfrac{x^2 + 4}{x + 1}$

(B) $\dfrac{(5x - 1)}{1} \cdot \dfrac{x - 1}{x^2 + 4}$

(C) $\dfrac{(5x - 1)}{1} \cdot \dfrac{x^2 + 4}{x - 1}$

(D) $\dfrac{1}{(5x - 1)} \cdot \dfrac{x + 1}{x^2 + 4}$

(E) $\dfrac{5x^2 - 1}{x - 1} \cdot \dfrac{x^2 + 4}{x - 1}$

**3. Multiple Choice**  What is the simplified form of the expression

$\dfrac{x + 3}{2x^2 + x - 10} \cdot \dfrac{2x + 5}{2x + 6}$?

(A) $\dfrac{1}{2(x - 2)}$  (B) $\dfrac{1}{2x + 5}$

(C) $\dfrac{x + 3}{2(x - 2)}$  (D) $\dfrac{2x + 5}{2(x + 3)}$

(E) $\dfrac{x + 3}{(x - 2)(2x + 6)}$

**4. Multiple Choice**  What is the simplified form of the expression $\dfrac{x^2 - 1}{x^2 + 6x} \div \dfrac{x - 1}{7x^2}$?

(A) $\dfrac{x - 1}{7x(x + 6)}$  (B) $\dfrac{7x(x + 1)}{x + 6}$

(C) $\dfrac{x + 6}{7(x + 1)}$  (D) $\dfrac{x + 1}{7(x + 6)}$

(E) $\dfrac{6x^2 - 1}{x^2 + 5x - 1}$

**5. Multiple Choice**  What is the simplified form of the expression

$\dfrac{5x^2 + 23x - 42}{x + 7} \div (5x - 7)$?

(A) $\dfrac{x + 7}{(x + 6)(5x - 7)}$  (B) $\dfrac{x + 7}{x + 6}$

(C) $\dfrac{5x - 7}{x + 6}$  (D) $\dfrac{x + 6}{x + 7}$

(E) $\dfrac{(x + 6)(5x - 7)^2}{x + 7}$

**6. Multiple Choice**  What is the simplified form of the expression

$\dfrac{x + 1}{15x^2 + 3x} \div \dfrac{x}{10x + 2}$?

(A) $\dfrac{x(x + 1)}{5x + 1}$  (B) $\dfrac{2x}{3(5x + 1)}$

(C) $\dfrac{2(x + 1)}{3x^2}$  (D) $\dfrac{2(5x + 1)}{3x^2}$

(E) $\dfrac{x + 1}{3x^2}$

**Quantitative Comparison**  In Exercises 7–9, simplify the expression if possible and evaluate it for $x = 4$. Then choose the statement below that is true about the given values.

(A)  The value in column A is greater.

(B)  The value in column B is greater.

(C)  The two values are equal.

(D)  The relationship cannot be determined from the given information.

| | Column A | Column B |
|---|---|---|
| 7. | $\dfrac{8x^3}{3} \cdot \dfrac{9}{2x}$ | $\dfrac{8x^3}{3} \div \dfrac{9}{2x}$ |
| 8. | $\dfrac{x^2 - 5x - 6}{x + 1} \div (x - 6)$ | $\dfrac{x^2 - 1}{x - 1} \cdot \dfrac{x + 1}{2}$ |
| 9. | $\dfrac{x^2 + 4x + 4}{x^2 + 5x + 4} \cdot \dfrac{x + 1}{x + 2}$ | $\dfrac{2x + 1}{x^2 + 8x} \div \dfrac{1}{x}$ |

**Algebra 1, Concepts and Skills**
Standardized Test Practice Workbook

# *Standardized Test Practice*

**For use with pages 658–662**

**TEST TAKING STRATEGY**  **Read all of the answer choices before deciding which is the correct one.**

**1.** *Multiple Choice*  Simplify the expression

$$\frac{7x + 2}{4x + 3} + \frac{6x}{4x + 3}.$$

  **(A)**   $x + 2$       **(B)** $\dfrac{x + 2}{4x + 3}$

  **(C)**   $\dfrac{13x + 2}{8x + 6}$     **(D)** $\dfrac{13x + 2}{4x + 3}$

  **(E)**   $\dfrac{42x^2 + 12x}{4x + 3}$

**2.** *Multiple Choice*  Simplify the expression

$$\frac{2x}{3x} - \frac{x}{3x}.$$

  **(A)**   $\dfrac{2}{3}$       **(B)** $\dfrac{2x}{3x}$

  **(C)**   $\dfrac{2}{3x}$      **(D)** $2x$

  **(E)**   None of these

**3.** *Multiple Choice*  Simplify the expression

$$\frac{x}{3x - 1} - \frac{3 - 8x}{3x - 1}.$$

  **(A)**   $1$       **(B)** $3$

  **(C)**   $\dfrac{7x - 3}{3x - 1}$     **(D)** $\dfrac{6x - 3}{3x - 1}$

  **(E)**   $\dfrac{7x + 3}{3x - 1}$

**4.** *Multiple Choice*  Simplify the expression

$$\frac{x^2 - 2}{x^2 + 3x + 2} + \frac{5x + 6}{x^2 + 3x + 2}.$$

  **(A)**   $\dfrac{x^2 + 5x + 4}{x^2 + 3x + 2}$    **(B)** $\dfrac{5x + 4}{3x + 2}$

  **(C)**   $\dfrac{x + 2}{x + 1}$      **(D)** $\dfrac{x + 4}{x + 2}$

  **(E)**   None of these

**5.** *Multiple Choice*  Simplify the expression

$$\frac{1}{(2x - 1)^2} - \frac{2x}{(2x - 1)^2}.$$

  **(A)**   $\dfrac{1 - 2x}{(2x - 1)^2}$    **(B)** $\dfrac{1 - 2x}{2x - 1}$

  **(C)**   $-\dfrac{1}{2x - 1}$     **(D)** $-1$

  **(E)**   $1 - 2x$

***Quantitative Comparison***  In Exercises 6–8, simplify the expression if possible and evaluate the expression for the given value of *x*. Then choose the statement below that is true about the result.

  **(A)**   The value in column A is greater.

  **(B)**   The value in column B is greater.

  **(C)**   The two values are equal.

  **(D)**   The relationship cannot be determined from the given information.

| | *Column A* | *Column B* |
|---|---|---|
| **6.** | $\dfrac{1}{x + 1} + \dfrac{x}{x + 1}$ for $x = 2$ | $\dfrac{1}{x - 1} + \dfrac{x}{x - 1}$ for $x = 2$ |
| **7.** | $\dfrac{y^2 - 2}{y - 1} + \dfrac{1}{y - 1}$ for $y = 0$ | $\dfrac{y^2 - 3y + 3}{y - 2} - \dfrac{y - 1}{y - 2}$ for $y = 0$ |
| **8.** | $\dfrac{y^2 - 2}{y - 1} + \dfrac{1}{y - 1}$ for $y = 4$ | $\dfrac{y^2 - 3y + 3}{y - 2} - \dfrac{y - 1}{y - 2}$ for $y = 4$ |

NAME _____ DATE _____

# *Standardized Test Practice*

**For use with pages 664–670**

**TEST TAKING STRATEGY**   **Read all of the answer choices before deciding which is the correct one.**

**1.** *Multiple Choice*   What is the LCD of $\dfrac{1}{5a}$, $\dfrac{a}{b}$, and $\dfrac{1 + a}{2 + a}$?

Ⓐ   $5a^2b + 2$        Ⓑ   $7 + a^2b$

Ⓒ   $7a^2b$              Ⓓ   $10a^2b$

Ⓔ   $5ab(2 + a)$

**2.** *Multiple Choice*   Find the LCD of $\dfrac{x}{x^2 - x - 6}$ and $\dfrac{3x + 1}{x^2 - 6x + 9}$.

Ⓐ   $(x + 2)(x - 3)$   Ⓑ   $(x - 3)(x - 3)$

Ⓒ   $x(3x + 1)$          Ⓓ   $(x - 2)(3x + 1)$

Ⓔ   $(x + 2)(x - 3)^2$

**3.** *Multiple Choice*   Simplify the expression $\dfrac{5}{x + 3} + \dfrac{x + 1}{x^2 + 5x + 6}$.

Ⓐ   $\dfrac{6x + 3}{(x + 2)(x + 3)}$        Ⓑ   $\dfrac{6x + 16}{(x + 2)(x + 3)}$

Ⓒ   $\dfrac{1}{x + 3}$                          Ⓓ   $\dfrac{6x + 11}{(x + 2)(x + 3)}$

Ⓔ   $\dfrac{6x^2 + 29x + 33}{(x + 2)^2(x + 3)^2}$

**4.** *Multiple Choice*   Simplify the expression $\dfrac{3x}{x - 1} - \dfrac{x - 1}{x + 5}$.

Ⓐ   $\dfrac{3x^2 + 15x}{x + 5}$              Ⓑ   $\dfrac{2x^2 + 17x - 1}{(x + 5)(x - 1)}$

Ⓒ   $\dfrac{3x^2 + 15x - 1}{x + 5}$          Ⓓ   $\dfrac{2x^2 - 2x + 16}{(x + 5)(x - 1)}$

Ⓔ   $\dfrac{2x + 1}{(x + 5)(x - 1)}$

**5.** *Multiple Choice*   Which expression represents the perimeter of the triangle?

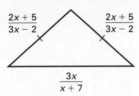

Ⓐ   $\dfrac{13x^2 + 32x + 70}{(3x - 2)(x + 7)}$

Ⓑ   $\dfrac{4x^2 + 42x + 68}{(3x - 2)(x + 7)}$

Ⓒ   $\dfrac{13x^2 + 14x + 25}{(3x - 2)(x + 7)}$

Ⓓ   $\dfrac{13x^2 + 44x + 70}{(x + 7)(3x - 2)}$

Ⓔ   $\dfrac{4x^2 + 42x + 68}{(3x - 2)^2(x + 7)}$

**6.** *Multi-Step Problem*   A plane is traveling 1200 miles between two cities. Its speed with no wind is represented by $x$ miles per hour. On a trip, traveling with the wind, the wind speed is 25 miles per hour. On the return trip, traveling against the wind, the wind speed increases to 30 miles per hour. Let $x + 25$ represent the speed with the wind and $x - 30$ represent the speed against the wind.

   **a.** Write an expression for the total time for the round trip.

   **b.** Simplify the expression from part (a).

   **c.** How long will the trip take, both ways, if the plane travels 200 miles per hour?

   **d.** *Critical Thinking*   A smaller plane can only travel 100 miles per hour. Compare the travel time of the planes. Did you get the result you expected? Explain.

NAME _____ DATE _____

# Standardized Test Practice

**For use with pages 670–677**

**TEST TAKING STRATEGY   As soon as the testing begins, start working. Keep a steady pace and stay focused on the test.**

1. *Multiple Choice*   What is the solution of
$\dfrac{x + 1}{5} = \dfrac{2x}{15}$?

   (A) $-\dfrac{3}{5}$      (B) $-3$      (C) $3$

   (D) $-\dfrac{1}{5}$      (E) $\dfrac{3}{5}$

2. *Multiple Choice*   What is a solution of
$\dfrac{6}{x + 2} = \dfrac{3x}{x^2 - 3x - 10}$?

   (A) $-2$           (B) $10$

   (C) $5$            (D) about $-4.4$

   (E) about $2.9$

3. *Multiple Choice*   You have a batting average of 0.200 after 90 times at bat. You would like to raise your average to 0.250. Which equation would you use to calculate the number of consecutive hits you need to achieve your goal?

   (A) $0.25 = \dfrac{18 + x}{90 + x}$   (B) $0.25 = \dfrac{18 + x}{90}$

   (C) $0.20 = \dfrac{18 + x}{90 + x}$   (D) $0.20 = \dfrac{x - 18}{90}$

   (E) $0.05 = \dfrac{x}{18}$

4. *Multiple Choice*   Two hoses are filling a swimming pool. The first hose can fill the pool in 30 minutes, the second in 45 minutes. If both hoses are used, how long will it take to fill the pool?

   (A) 15 minutes      (B) 16 minutes

   (C) 18 minutes      (D) 20 minutes

   (E) 22 minutes

5. *Multiple Choice*   You make a mixture of nuts costing $5 per pound and chocolate drops costing $7.50 per pound. How many pounds of nuts do you need to make 10 pounds of a mixture worth $6 per pound?

   (A) 3 lb      (B) 7 lb      (C) 5 lb

   (D) 4 lb      (E) 6 lb

6. *Multiple Choice*   What is the solution of
$\dfrac{x}{x + 1} = 2 + \dfrac{5}{x + 1}$?

   (A) $-1$      (B) $7$      (C) $5$

   (D) $-5$      (E) $-7$

*Quantitative Comparison*   In Exercises 7-9, solve the equation. Then choose the statement below that is true about the solutions.

   (A) The solution in column A is greater.

   (B) The solution in column B is greater.

   (C) The two solutions are equal.

   (D) The relationship cannot be determined from the given information.

| | Column A | Column B |
|---|---|---|
| 7. | $\dfrac{5}{2x - 1} = \dfrac{1}{x}$ | $\dfrac{x}{x + 1} = \dfrac{1}{2}$ |
| 8. | $\dfrac{1}{x} + \dfrac{1}{2} = \dfrac{5}{x}$ | $\dfrac{1}{x + 2} - \dfrac{1}{2} = 2$ |
| 9. | $\dfrac{1}{y - 5} - \dfrac{3}{y + 5} = 0$ | $\dfrac{2}{3 - y} = \dfrac{5}{2y + 1}$ |

**Algebra 1, Concepts and Skills**
Standardized Test Practice Workbook

# *Standardized Test Practice*

**For use with pages 692–697**

**TEST TAKING STRATEGY**   **If you can, check your answer using a different method than you used originally, to avoid making the same mistake twice.**

1. *Multiple Choice*   What is the domain of the function $y = \sqrt{x + 6}$?

   Ⓐ $x \geq -6$   Ⓑ $x \leq -6$   Ⓒ $x \geq 6$

   Ⓓ $x \leq 6$   Ⓔ $x \geq 0$

2. *Multiple Choice*   Which quadrants of the coordinate plane will contain the graph of $y = \sqrt{2x - 4}$?

   Ⓐ Quadrant I   Ⓑ Quadrant III

   Ⓒ Quadrant IV   Ⓓ Quadrant I and IV

   Ⓔ Quadrant III and IV

3. *Multiple Choice*   What are the domain and the range of the function $y = \sqrt{x - 2}$?

   Ⓐ Domain: $x \geq 0$   Range: $x \geq 0$

   Ⓑ Domain: $x \geq 2$   Range: $x \geq 0$

   Ⓒ Domain: $x \geq 0$   Range: $y \geq -2$

   Ⓓ Domain: $x \geq -2$   Range: $y \geq 2$

   Ⓔ Domain: $x \geq 0$   Range: $y \geq 2$

4. *Multiple Choice*   What is the equation of the graph of the function shown?

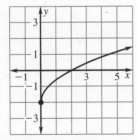

   Ⓐ $y = \sqrt{2x} - 2$   Ⓑ $y = \sqrt{x} - 2$

   Ⓒ $y = \sqrt{x} + 2$   Ⓓ $y = \sqrt{2x - 2}$

   Ⓔ $y = \sqrt{x - 2}$

5. *Multiple Choice*   What is the maximum walking speed of an animal with a leg length of 3 feet? The walking speed model is $S = \sqrt{32L}$, where $S$ is the speed in feet per second, and $L$ is the leg length in feet.

   Ⓐ about 4.8 ft/sec   Ⓑ about 3.3 ft/sec

   Ⓒ about 9.8 ft/sec   Ⓓ about 5.9 ft/sec

   Ⓔ about 6.6 ft/sec

6. *Multiple Choice*   What is the radius of a cylindrical container with a volume of 196.25 cubic inches? The height is 10 inches. (*Hint:* $V = \pi r^2 h$.)

   Ⓐ about 2.2 inches   Ⓑ about 10.3 inches

   Ⓒ about 2.5 inches   Ⓓ about 13.5 inches

   Ⓔ about 12.8 inches

*Quantitive Comparison*   In Exercises 7–9, evaluate each square root function for the given value of $x$, then choose the statement that is true about the given numbers.

   Ⓐ The value in column A is greater.

   Ⓑ The value in column B is greater.

   Ⓒ The two values are equal.

   Ⓓ The relationship cannot be determined from the given information.

|   |   | *Column A* | *Column B* |
|---|---|---|---|
| 7. | When $x = 3$ | $y = \dfrac{1}{3}\sqrt{3x} - 1$ | $y = \sqrt{6 - 2x}$ |
| 8. | When $x = -2$ | $y = \sqrt{6 - 5x}$ | $y = \sqrt{4x^2 + 9}$ |
| 9. | When $x = 5$ | $y = \sqrt{\dfrac{x}{4} - 1}$ | $y = \sqrt{\dfrac{x}{3} + 1}$ |

**Algebra 1, Concepts and Skills**
Standardized Test Practice Workbook

**TEST TAKING STRATEGY** **Work as fast as you can through the easier problems, but not so fast that you are careless.**

**1.** *Multiple Choice*  Simplify the expression $\sqrt{20} - \sqrt{80}$.

   Ⓐ $2\sqrt{5}$    Ⓑ $-12\sqrt{5}$  Ⓒ $12\sqrt{5}$

   Ⓓ $-2\sqrt{5}$  Ⓔ $-\sqrt{20}$

**2.** *Multiple Choice*  Simplify the expression $\dfrac{1}{\sqrt{2}}$.

   Ⓐ $\dfrac{2}{\sqrt{2}}$    Ⓑ $\sqrt{2}$    Ⓒ $2$

   Ⓓ $\dfrac{\sqrt{2}}{2}$   Ⓔ None of these

**3.** *Multiple Choice*  Simplify the expression $\dfrac{1}{\sqrt{3} - 2}$.

   Ⓐ $\sqrt{3} - 2$   Ⓑ $-\dfrac{\sqrt{3} + 2}{2}$

   Ⓒ $\sqrt{3} + 2$   Ⓓ $-\sqrt{3} + 2$

   Ⓔ $-\sqrt{3} - 2$

**4.** *Multiple Choice*  Simplify $\sqrt{3}\,(2\sqrt{3} - \sqrt{2})$.

   Ⓐ $6 - 3\sqrt{2}$   Ⓑ $6 - \sqrt{6}$

   Ⓒ $3\sqrt{3} - 3\sqrt{2}$ Ⓓ $6\sqrt{3} - 3\sqrt{2}$

   Ⓔ $6\sqrt{3} - \sqrt{6}$

**5.** *Multiple Choice*  Find the area of the triangle below.

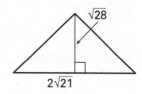

   Ⓐ $28\sqrt{3}$

   Ⓑ $7$

   Ⓒ $14\sqrt{3}$

   Ⓓ $7\sqrt{3}$

   Ⓔ $2\sqrt{21}$

**6** *Multiple Choice*  One pole-vaulter reaches a height of 16 feet and another reaches a height of 12 feet. Use the equation $V = 8\sqrt{h}$, where $V$ is velocity and $h$ is height, to find the difference in their velocities in feet per second.

   Ⓐ $32$          Ⓑ $512\sqrt{3}$

   Ⓒ $32 - 32\sqrt{3}$  Ⓓ $32 + 16\sqrt{3}$

   Ⓔ $32 - 16\sqrt{3}$

*Quantitative Comparison*  In Exercises 7–9, simplify the expressions. Then choose the statement below that is true about the given numbers.

   Ⓐ The value in column A is greater.

   Ⓑ The value in column B is greater.

   Ⓒ The two values are equal.

   Ⓓ The relationship cannot be determined from the given information.

|  | Column A | Column B |
|---|---|---|
| **7.** | $(2 + \sqrt{6})(2 - \sqrt{6})$ | $2\sqrt{3} \cdot \sqrt{12}$ |
| **8.** | $\sqrt{75} - \sqrt{27}$ | $\sqrt{72} - \sqrt{18}$ |
| **9.** | $\dfrac{3}{\sqrt{5}} \cdot \dfrac{2\sqrt{15}}{3}$ | $\dfrac{\sqrt{12}}{3} \cdot \dfrac{\sqrt{18}}{\sqrt{2}}$ |

*Chapter 12*

NAME _____ DATE _____

# *Standardized Test Practice*

For use with pages 704–709

**TEST TAKING STRATEGY** **Learn as much as you can about a test ahead of time, such as the types of questions and the topics that the test will cover.**

1. *Multiple Choice* Which of the following is a solution of $\sqrt{x} - 5 = 10$?
   - Ⓐ 15
   - Ⓑ 30
   - Ⓒ 5.5
   - Ⓓ 225
   - Ⓔ 230

2. *Multiple Choice* Which of the following is a solution of $\sqrt{2x + 3} - 3 = 8$?
   - Ⓐ 118
   - Ⓑ 62
   - Ⓒ 59
   - Ⓓ 32
   - Ⓔ 31

3. *Multiple Choice* Which of the following is a solution of $3 - \sqrt{2x - 1} = 5$?
   - Ⓐ $\frac{2}{5}$
   - Ⓑ $\frac{5}{2}$
   - Ⓒ $\frac{3}{5}$
   - Ⓓ $\frac{5}{3}$
   - Ⓔ no solution

4. *Multiple Choice* Which of the following is a solution of $x = \sqrt{4x + 32}$?
   - Ⓐ 8
   - Ⓑ 4
   - Ⓒ −4
   - Ⓓ 8, 4
   - Ⓔ 8, −4

5. *Multiple Choice* Which of the following is a solution of $x = \sqrt{2x + 48}$?
   - Ⓐ 6
   - Ⓑ 8
   - Ⓒ −24
   - Ⓓ −6
   - Ⓔ −8

6. *Multiple Choice* Which of the following is a solution of $\sqrt{5 - 2x} - 1 = 0$?
   - Ⓐ 2
   - Ⓑ −2
   - Ⓒ 3
   - Ⓓ −3
   - Ⓔ 0

7. *Multiple Choice* Which of the following is a solution of $\sqrt{2x + 3} = x$?
   - Ⓐ −3, 1
   - Ⓑ −1
   - Ⓒ 3
   - Ⓓ 3, −1
   - Ⓔ 1

8. *Multiple Choice* What is the value of $x$ in the triangle below whose area is 26?
   - Ⓐ 1
   - Ⓑ 2
   - Ⓒ 3
   - Ⓓ 4
   - Ⓔ 5

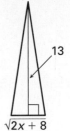

13

$\sqrt{2x + 8}$

9. *Multiple Choice* The square root of the product of two positive numbers is 6. One number is 9 less than the other. What are the numbers?
   - Ⓐ 4, 8
   - Ⓑ 3, 12
   - Ⓒ 3, 6
   - Ⓓ 6, 12
   - Ⓔ 6, 8

*Quantitative Comparison* In Exercises 10–12, choose the statement below that is true about the given quantities.

- Ⓐ The quantity in column A is greater.
- Ⓑ The quantity in column B is greater.
- Ⓒ The two quantities are equal.
- Ⓓ The relationship cannot be determined from the given information.

|  | Column A | Column B |
|---|---|---|
| **10.** | The solution of $\sqrt{x - 1} = 0$ | The solution of $\sqrt{x} = 1$ |
| **11.** | The solution of $\sqrt{x + 3} = 5$ | The solution of $\sqrt{x + 5} = 3$ |
| **12.** | The solution of $\sqrt{x + 1} = 1$ | The solution of $\sqrt{x + 2} = 2$ |

**Algebra 1, Concepts and Skills**
Standardized Test Practice Workbook

NAME _____ DATE _____

# Standardized Test Practice

For use with pages 710–714

**TEST TAKING STRATEGY**   **Check your work if you have time, but do not change your answers unless you are sure they are wrong.**

1. **Multiple Choice**   Rewrite the expression $5^{\frac{2}{5}}$ using radical notation.

   (A) $(\sqrt{5})^5$   (B) $\sqrt[5]{5}$   (C) $(\sqrt[5]{5})^2$

   (D) $(\sqrt[2]{2})^5$   (E) $(\sqrt{2})^2$

2. **Multiple Choice**   Evaluate $4^{\frac{3}{2}}$ without using a calculator.

   (A) 2   (B) 4   (C) 8

   (D) 16   (E) 32

3. **Multiple Choice**   Evaluate $(\sqrt[3]{27})^4$ without using a calculator.

   (A) 20.25   (B) 49   (C) 54

   (D) 81   (E) 144

4. **Multiple Choice**   Evaluate $81^{\frac{3}{4}}$ without using a calculator.

   (A) 27   (B) 60.75   (C) 9

   (D) 18   (E) 54

5. **Multiple Choice**   Rewrite the expression $(\sqrt[4]{5})^7$ using rational exponent notation.

   (A) $5^{\frac{4}{7}}$   (B) $5^{\frac{7}{4}}$   (C) $4^{\frac{5}{7}}$

   (D) $4^{\frac{7}{5}}$   (E) $7^{\frac{4}{5}}$

6. **Multiple Choice**   Simplify $5^{\frac{7}{8}} \cdot 5^{\frac{1}{4}}$.

   (A) $5^{\frac{2}{3}}$   (B) $25^{\frac{9}{8}}$   (C) $25^{\frac{7}{32}}$

   (D) $5^{\frac{7}{32}}$   (E) $5^{\frac{9}{8}}$

7. **Multiple Choice**   Simplify $\left(12^{\frac{1}{2}}\right)^{\frac{1}{4}}$.

   (A) $12^{\frac{1}{6}}$   (B) $12^{\frac{1}{8}}$   (C) $12^{\frac{1}{3}}$

   (D) 4   (E) $12^{\frac{1}{4}}$

8. **Multiple Choice**   Simplify $\left(x^{\frac{1}{3}} \cdot x^{\frac{1}{5}}\right)^{\frac{1}{2}}$.

   (A) $x^{\frac{1}{16}}$   (B) $x^{\frac{9}{17}}$   (C) $x^{\frac{8}{30}}$

   (D) $x^{\frac{4}{15}}$   (E) $x^{\frac{31}{30}}$

**Quantitative Comparison**   In Exercises 9–11, evaluate the expressions. Then choose the statement below that is true about the given numbers.

   (A) The value in column A is greater.

   (B) The value in column B is greater.

   (C) The values are equal.

   (D) The relationship cannot be determined from the given information.

| | Column A | Column B |
|---|---|---|
| 9. | $9^{\frac{3}{2}}$ | $8^{\frac{4}{3}}$ |
| 10. | $\left(\frac{1}{4}\right)^{\frac{5}{2}}$ | $\left(\frac{1}{9}\right)^{\frac{3}{2}}$ |
| 11. | $81^{\frac{3}{4}}$ | $125^{\frac{2}{3}}$ |

**Algebra 1, Concepts and Skills**
Standardized Test Practice Workbook

Chapter 12

## *Standardized Test Practice*

For use with pages 716–721

**TEST TAKING STRATEGY   Avoid spending too much time on one question. Skip questions that are too difficult for you, and spend no more than a few minutes on each question.**

**1.** *Multiple Choice*   What term should be added to $x^2 + 18x$ so that the result is a perfect square trinomial?

  Ⓐ 9      Ⓑ 162      Ⓒ 42

  Ⓓ 81      Ⓔ 36

**2.** *Multiple Choice*   What term should be added to $x^2 - \frac{6}{5}x$ so that the result is a perfect square trinomial?

  Ⓐ $\frac{3}{5}$      Ⓑ $\frac{144}{25}$      Ⓒ $\frac{9}{25}$

  Ⓓ $\frac{12}{5}$      Ⓔ $-\frac{3}{5}$

**3.** *Multiple Choice*   Solve $x^2 - 4x - 8 = 0$ by completing the square, then choose the solution.

  Ⓐ $2 \pm 2\sqrt{2}$      Ⓑ $2 \pm 2\sqrt{3}$

  Ⓒ $-2 \pm 2\sqrt{2}$      Ⓓ $-2 \pm 2\sqrt{3}$

  Ⓔ None of these

**4.** *Multiple Choice*   Solve $x^2 - 12x - 7 = 0$ by completing the square.

  Ⓐ $7 \pm \sqrt{43}$   Ⓑ $6 \pm \sqrt{43}$

  Ⓒ $6 \pm \sqrt{42}$   Ⓓ $7 \pm \sqrt{42}$

  Ⓔ $12 \pm \sqrt{43}$

**5.** *Multiple Choice*   Solve $15x - 45x^2 = 0$ using the most appropriate method.

  Ⓐ $0, -\frac{1}{3}$      Ⓑ $0, 3$      Ⓒ $0, \frac{1}{3}$

  Ⓓ $1, \frac{1}{3}$      Ⓔ $0, -3$

**6.** *Multiple Choice*   Solve the equation $3x^2 + 13x - 10 = 0$ using the most appropriate method.

  Ⓐ $-\frac{2}{3}, 5$      Ⓑ $-5, \frac{2}{3}$      Ⓒ $-5, \frac{3}{2}$

  Ⓓ $10, \frac{5}{3}$      Ⓔ $-10, \frac{5}{3}$

**7.** *Multiple Choice*   The area of a triangle is 90 square inches. The height is $x$ inches and the base is $(4x - 8)$ inches. What is the height?

  Ⓐ 5.8 inches      Ⓑ 7.0 inches

  Ⓒ 8.2 inches      Ⓓ 7.8 inches

  Ⓔ 4.3 inches

**8.** *Multiple Choice*   Solve $x^2 - 14x - 11 = 0$ using the most appropriate method.

  Ⓐ $7 \pm 6\sqrt{10}$      Ⓑ $-7 \pm 3\sqrt{10}$

  Ⓒ $-7 \pm 2\sqrt{5}$      Ⓓ $-7 \pm 3\sqrt{5}$

  Ⓔ $7 \pm 2\sqrt{15}$

*Quantitative Comparison*   In Exercises 9–11, find the term that should be added to the expression to create a perfect square trinomial. Then choose the statement below that is true about the added terms.

  Ⓐ The added term in column A is greater.

  Ⓑ The added term in column B is greater.

  Ⓒ The two added terms are equal.

  Ⓓ The relationship cannot be determined from the given information.

| | Column A | Column B |
|---|---|---|
| **9.** | $x^2 + 13x$ | $x^2 - 13x$ |
| **10.** | $x^2 + \frac{3}{5}x$ | $x^2 - \frac{4}{7}x$ |
| **11.** | $x^2 + 3.2x$ | $x^2 + \frac{17}{5}x$ |

*Chapter 12*

## *Standardized Test Practice*

For use with pages 722–729

**TEST TAKING STRATEGY**   Be aware of how much time you have left, but keep focused on your work.

1. *Multiple Choice*   What is the unknown length of the right triangle?

   Ⓐ  $6\sqrt{3}$   Ⓑ  $18\sqrt{2}$

   Ⓒ  $3\sqrt{6}$   Ⓓ  $4\sqrt{3}$

   Ⓔ  $3\sqrt{4}$

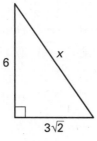

2. *Multiple Choice*   What is the unknown length of the right triangle?

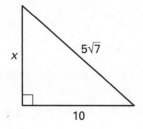

   Ⓐ  $5\sqrt{3}$   Ⓑ  $7\sqrt{2}$

   Ⓒ  $11\sqrt{5}$   Ⓓ  $3\sqrt{5}$

   Ⓔ  $5\sqrt{11}$

3. *Multiple Choice*   A right triangle has one leg that is 5 inches longer than the other leg. The hypotenuse is 25. Find the length of the shorter leg.

   Ⓐ  10     Ⓑ  15     Ⓒ  17

   Ⓓ  20     Ⓔ  13

4. *Multiple Choice*   A right triangle has one leg that is twice as long as the other leg. If the hypotenuse is $8\sqrt{5}$, find the length of the longer leg.

   Ⓐ  8     Ⓑ  16     Ⓒ  4

   Ⓓ  32     Ⓔ  12

5. *Multiple Choice*   Which lengths below would form a right triangle?

   Ⓐ  1, 3, 5     Ⓑ  1.2, 2.4, 6.1

   Ⓒ  6, 9, 10     Ⓓ  $2\sqrt{5}$, 4, 6

   Ⓔ  None of these

6. *Multiple Choice*   A rectangular pool is 13 feet by 25 feet. What is the diagonal length from corner to corner?

   Ⓐ  about 29.5   Ⓑ  about 27 ft

   Ⓒ  about 27.5   Ⓓ  about 28.5

   Ⓔ  about 28 ft

7. *Multiple Choice*   The area of a square is 64 square feet. What is the length of its diagonal?

   Ⓐ  8 feet     Ⓑ  $8\sqrt{3}$ feet

   Ⓒ  $9\sqrt{2}$ feet     Ⓓ  $8\sqrt{2}$ feet

   Ⓔ  $5\sqrt{6}$ feet

8. *Multiple Choice*   The diagonal of a square is $5\sqrt{2}$ feet. What is its area?

   Ⓐ  5 ft$^2$     Ⓑ  10 ft$^2$

   Ⓒ  15 ft$^2$     Ⓓ  25 ft$^2$

   Ⓔ  50 ft$^2$

9. *Multi-Step Problem*   Kim and Cindy bike from the same point at the same time. Cindy travels west and Kim travels south. After one hour, Cindy has traveled 2 miles farther than Kim, and they are 10 miles apart.

   a. Let $d$ represent Kim's distance. Write an expression for Cindy's distance.

   b. Draw a diagram of the situation.

   c. Use the Pythagorean theorem to find the distance each person traveled.

   d. *Writing*   Which method did you use to solve the quadratic equation? Why?

**TEST TAKING STRATEGY**  **Think positively during a test. This will help keep up your confidence and enable you to focus on each question.**

1. *Multiple Choice*  What is the distance between $(-3, -2)$ and $(-4, 3)$?

   **A** $\sqrt{74}$   **B** $\sqrt{2}$   **C** $\sqrt{34}$

   **D** $\sqrt{26}$   **E** $\sqrt{8}$

2. *Multiple Choice*  What is the distance between $(3, 5)$ and $(8, -7)$?

   **A** $\sqrt{29}$   **B** $13$   **C** $\sqrt{109}$

   **D** $5$   **E** $\sqrt{265}$

3. *Multiple Choice*  What is the distance between between points $A$ and $B$?

   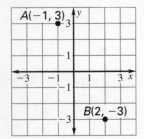

   **A** $3\sqrt{5}$   **B** $3$   **C** $\sqrt{10}$

   **D** $3\sqrt{10}$   **E** $\sqrt{5}$

4. *Multiple Choice*  The vertices of a right triangle are $(5, -2)$, $(0, 6)$, and $(0, -2)$. What is the length of the hypotenuse?

   **A** $8$   **B** $4$   **C** $\sqrt{89}$

   **D** $\sqrt{41}$   **E** $5$

5. *Multiple Choice*  The vertices of a right triangle are $(-1, -5)$, $(3, 2)$, and $(3, -5)$. What is the length of the hypotenuse?

   **A** $4$   **B** $7$   **C** $\sqrt{65}$

   **D** $\sqrt{13}$   **E** $3\sqrt{5}$

6. *Multiple Choice*  Which set of points below form a right triangle?

   **A** $(3, 2), (5, 7), (-1, 7)$

   **B** $(-2, 1), (4, 8), (6, 1)$

   **C** $(-1, 2), (1, 2), (6, -1)$

   **D** $(-5, -3), (4, -3), (-5, 4)$

   **E** None of these

7. *Multiple Choice*  Two fish are stocked into a lake from a boat. The first swims 1 mile north and 2 miles east. The second swims 3 miles west then 2 miles north. How many miles apart are the fish?

   **A** $5.8$   **B** $4$   **C** $4.5$

   **D** $5.1$   **E** $5.6$

*Quantitative Comparison*  In Exercises 8–10, find the distance between the two points. Then choose the statement below that is true about the given numbers.

   **A**  The value in column A is greater.

   **B**  The value in column B is greater.

   **C**  The two values are equal.

   **D**  The relationship cannot be determined from the given information.

| | Column A | Column B |
|---|---|---|
| **8.** | $(5, 8)$ and $(2, 1)$ | $(-4, 2)$ and $(3, 2)$ |
| **9.** | $(-2, -2)$ and $(-5, -13)$ | $(4, -2)$ and $(7, 9)$ |
| **10.** | $\left(\frac{1}{2}, 6\right)$ and $\left(2, \frac{2}{3}\right)$ | $\left(\frac{5}{4}, -2\right)$ and $\left(3, -\frac{1}{2}\right)$ |

Chapter 12

NAME _____ DATE _____

# Standardized Test Practice

For use with pages 736–739

TEST TAKING STRATEGY   **Answer the questions you are sure of first. If you cannot solve a problem, move on to the next question; come back to it later if you have time.**

**1.** *Multiple Choice*   What is the midpoint between $(0, -2)$ and $(-4, 3)$?

(A)   $(3, 6)$     (B)   $(6, 3)$

(C)   $(0.5, 2)$     (D)   $(-3, 0.5)$

(E)   $(-2, 0.5)$

**2.** *Multiple Choice*   What is the midpoint between $(-7, 2)$ and $(7, 2)$?

(A)   $(-7, 0)$     (B)   $(0, 2)$

(C)   $(-7, 2)$     (D)   $(7, 2)$

(E)   $(0, 7)$

**3.** *Multiple Choice*   What is the midpoint between points $A$ and $B$?

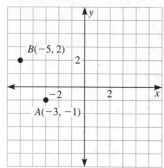

(A)   $(-4, 0.5)$     (B)   $(-4, -0.5)$

(C)   $(0.5, 4)$     (D)   $(-0.5, -4)$

(E)   None of these

**4.** *Multiple Choice*   What is the midpoint between $(10, 3)$ and $(-6, 2)$?

(A)   $\left(8, -\frac{5}{2}\right)$     (B)   $\left(2, \frac{5}{2}\right)$

(C)   $\left(8, -\frac{1}{2}\right)$     (D)   $\left(-2, -\frac{5}{2}\right)$

(E)   $\left(-2, -\frac{1}{2}\right)$

**5.** *Multiple Choice*   The midpoint between two points is $(3, -4)$ and the points are $(5, 2)$ and $(x, y)$. Find the point $(x, y)$.

(A)   $(1, 10)$     (B)   $(-2, 10)$

(C)   $(1, -10)$     (D)   $(2, 10)$

(E)   $(-1, -2)$

**6.** *Multiple Choice*   What is the midpoint between $(0, 0)$ and $\left(-45, \frac{5}{8}\right)$?

(A)   $\left(22.5, -\frac{5}{16}\right)$     (B)   $\left(22.5, \frac{5}{4}\right)$

(C)   $\left(-22.5, -\frac{5}{4}\right)$     (D)   $\left(-22.5, \frac{5}{16}\right)$

(E)   $\left(-22.5, -\frac{5}{16}\right)$

**7.** *Multi-Step Problem*   You are given the following two points:
$$A(-3, 5) \text{ and } B(2, -7).$$

**a.** Find the midpoint between the two given points.

**b.** Find the distance between point $A$ and the midpoint.

**c.** Find the distance between point $B$ and the midpoint.

**d.** Find the distance from point $A$ to point $B$. Compare this to the sum of the results from parts (b) and (c).

# *Standardized Test Practice*

For use with pages 740–746

**TEST TAKING STRATEGY**   **As soon as the testing begins, start working. Keep a steady pace and stay focused on the test.**

1. *Multiple Choice*   The basic axiom of algebra represented by $x\left(\dfrac{1}{x}\right) = 1$ where $x$ is any real number not equal to zero is the ___?___.

   Ⓐ   commutative property of multiplication
   Ⓑ   associative property of multiplication
   Ⓒ   identity property of multiplication
   Ⓓ   inverse property of multiplication
   Ⓔ   substitution property of equality

2. *Multiple Choice*   The basic axiom of algebra represented by $7(x + y) = 7x + 7y$, where $x$ and $y$ are real numbers, is the ___?___.
   Ⓐ   closure property of multiplication
   Ⓑ   commutative property of multiplication
   Ⓒ   associative property of multiplication
   Ⓓ   distributive property
   Ⓔ   associative property of addition

3. *Multiple Choice*   The basic axiom of algebra represented by $(12x)y = 12(xy)$, where $x$ and $y$ are real numbers, is the ___?___.
   Ⓐ   closure property of multiplication
   Ⓑ   commutative property of multiplication
   Ⓒ   associative property of multiplication
   Ⓓ   distributive property
   Ⓔ   identity property of multiplication

4. *Multiple Choice*   The basic axiom of algebra represented by $(1 + 2) + 3 = 1 + (2 + 3)$ is the ___?___.
   Ⓐ   closure property of addition
   Ⓑ   inverse property of addition
   Ⓒ   identity property of addition
   Ⓓ   commutative property of addition
   Ⓔ   associative property of addition

*Multiple Choice*   In Exercises 5–8, choose one of items A–E as a justification for the step taken in the proof below. You may use a choice more than once.

   Ⓐ   Definition of subtraction
   Ⓑ   Distributive property
   Ⓒ   Associative property of multiplication
   Ⓓ   For all real numbers $b$ and $c$,
        $c(-b) = -cb$.
   Ⓔ   Identity property of multiplication

   If $x$ and $y$ are real numbers, then
   $$5(x - y) = 5x - 5y.$$

5.   $5(x - y) = 5[x + (-y)]$   ___?___

6.   $= 5x + 5(-y)$   ___?___

7.   $= 5x + (-5y)$   ___?___

8.   $= 5x - 5y$   ___?___

9. *Multi-Step Problem*   In parts (*a.*) and (*b.*), give a counterexample to show that each statement is *not* true.

   a.   $(a - b) - c = a - (b - c)$.

   b.   If $a$ is a positive integer, then $a^2 > a$.

   c.   Find values of $a$, $b$, and $c$ for which the statement in part (a) is true.

   d.   Find a value of $a$ for which the statement in part (b) is true.

# Cumulative Standardized Test Practice

**For use after Chapters 1–12**

1. **Multiple Choice**  Evaluate the expression $x^2 - 3y^2 + 8$ when $x = 2$ and $y = -2$.

   **A**  12      **B**  0      **C**  $-12$

   **D**  24      **E**  $-24$

2. **Multiple Choice**  Simplify $3x(x - 2) - 2x + x^2$.

   **A**  $2x^2 - 8x$      **B**  $4x^2 - 4x$

   **C**  $4x^2 - 2x - 2$      **D**  $4x^2 + 8x$

   **E**  $4x^2 - 8x$

**Quantitative Comparison**  In Exercises 3–6, choose the statement below that is true about the given quantities.

   **A**  The value in column A is greater.

   **B**  The value in column B is greater.

   **C**  The two values are equal.

   **D**  The relationship cannot be determined from the given information.

| | Column A | Column B |
|---|---|---|
| **3.** | $x$ when $12x - 3 = 4$ | $x$ when $2x - 5 = 3x - 8$ |
| **4.** | $x$ when $\frac{1}{2}x + |-3| = 7$ | $x$ when $7x + 2(-x + 6) = 3$ |
| **5.** | The slope of $y = \frac{1}{2}x - \frac{1}{2}$ | The $y$-intercept of $y = \frac{1}{2}x - \frac{1}{2}$ |
| **6.** | $x$ when $-15 = \frac{x}{5}$ | $x$ when $\frac{1}{5}x = -15$ |

7. **Multiple Choice**  What is the length of a rectangular container if its volume is 336 cubic inches? Its height is 6 inches and its width is 8 inches. (*Hint: V = lwh*)

   **A**  6 in.      **B**  8 in.      **C**  5 in.

   **D**  7 in.      **E**  3 in.

8. **Multiple Choice**  Find the slope of the line passing through the points $(5, -2)$ and $(11, 2)$.

   **A**  0      **B**  $\frac{2}{3}$      **C**  $-\frac{2}{3}$

   **D**  $\frac{3}{2}$      **E**  $-\frac{3}{2}$

9. **Multiple Choice**  What is an equation of a line perpendicular to $y = \frac{1}{3}x - 5$?

   **A**  $y = 3x + 6$      **B**  $y = \frac{1}{3}x + \frac{1}{5}$

   **C**  $y = -\frac{1}{3}x + 2$      **D**  $y = -3x - 8$

   **E**  $y = -\frac{1}{3}x + 5$

10. **Multiple Choice**  What is the standard form of an equation that passes through the points $(14, 8)$ and $(3, 11)$?

   **A**  $3x + 11y = 118$

   **B**  $3x - 11y = 130$

   **C**  $3x + 11y = -130$

   **D**  $3x - 11y = -118$

   **E**  $3x + 11y = 130$

11. **Multiple Choice**  What is the solution of the inequality $8 \le -3x + 1 \le 11$?

   **A**  $-3 \le x \le -4$      **B**  $-3 \ge x \ge -4$

   **C**  $3 \ge x \ge 4$      **D**  $-\frac{10}{3} \ge x \ge -\frac{7}{3}$

   **E**  $-\frac{10}{3} \le x \le -\frac{7}{3}$

12. **Multiple Choice**  If $y = 4x - 2$ and $7x - 2y = 1$, then $x = \underline{\ ?\ }$.

   **A**  1      **B**  3      **C**  10

   **D**  0      **E**  5

**Chapter 12**

**13. *Multiple Choice*** Use the substitution method to solve the system of linear equations.

$$3x + 4y = 14$$
$$y = -\frac{5}{2}x$$

Ⓐ $(-1, 8)$    Ⓑ $(4, -10)$

Ⓒ $(-2, 5)$    Ⓓ $(2, 2)$    Ⓔ $\left(\frac{2}{3}, 3\right)$

**14. *Multiple Choice*** Solve the system of linear equations using linear combinations.

$$5x - 3y = 26$$
$$3x + 2y = 8$$

Ⓐ $(-2, 4)$    Ⓑ $(7, 3)$    Ⓒ $(2, 4)$

Ⓓ $(4, -2)$    Ⓔ $(0, 4)$

**15. *Multiple Choice*** Choose the system of linear inequalities represented by the graph.

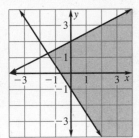

Ⓐ $y \le \frac{1}{2}x + 2$    Ⓑ $y \le \frac{1}{2}x + 2$

    $y \ge -\frac{3}{2}x - 1$     $y \le -\frac{3}{2}x - 1$

Ⓒ $y \ge \frac{1}{2}x + 2$    Ⓓ $y \ge \frac{1}{2}x + 2$

    $y \ge -\frac{3}{2}x - 1$     $y \le -\frac{3}{2}x - 1$

Ⓔ None of these

**16. *Multiple Choice*** Simplify the expression

$$\frac{5x^2y^{-3}}{8x^3y} \cdot \frac{4x^{-2}y^3}{10xy^{-4}}.$$

Ⓐ $\frac{x^2y^3}{4}$    Ⓑ $\frac{y^3}{4x^2}$    Ⓒ $\frac{y}{4x^4}$

Ⓓ $4y^3x^4$    Ⓔ $\frac{y^3}{4x^4}$

***Quantitative Comparison*** In Exercises 17–19, choose the statement that is true about the given quantities.

Ⓐ The value of column A is greater.

Ⓑ The value of column B is greater.

Ⓒ The two values are equal.

Ⓓ The relationship cannot be determined from the information given.

| | Column A | Column B |
|---|---|---|
| **17.** | $(2x)^2(x^2)^3$ when $x = 2$ | $(7x^3)^2$ when $x = 2$ |
| **18.** | $\left(\frac{3}{2}\right)^{-3}$ | $\left(\frac{1}{2}\right)^2\left(\frac{1}{3}\right)^{-2}$ |
| **19.** | $y = 14\left(\frac{9}{7}\right)^t$ when $t = 2$ | $y = 6(1.5)^t$ when $t = 2$ |

**20. *Multiple Choice*** You deposit $1200 into a savings account that pays 7% interest compounded yearly. How much money is in the account after 8 years assuming you made no additional investments or withdrawals?

Ⓐ $1872    Ⓑ $2056.59

Ⓒ $2117.86    Ⓓ $2061.82

Ⓔ $8370.91

**21. *Multiple Choice*** Which is the simplified form of $\sqrt{\dfrac{75}{6}}$?

Ⓐ $\dfrac{5\sqrt{3}}{2}$    Ⓑ $\dfrac{5}{\sqrt{2}}$

Ⓒ $\dfrac{5\sqrt{2}}{2}$    Ⓓ $\dfrac{5\sqrt{3}}{6}$

Ⓔ $\dfrac{5\sqrt{2}}{6}$

NAME _____ DATE _____

## *Cumulative Standardized Test Practice*

For use after Chapters 1–12

---

**22.** *Multiple Choice* Evaluate $\sqrt{b^2 - 4ac}$ when $a = 7, b = -2, c = -6$.

- **A** $2\sqrt{51}$
- **B** $\sqrt{97}$
- **C** $2\sqrt{43}$
- **D** $2\sqrt{23}$
- **E** undefined

**23.** *Multiple Choice* What is the vertex of the graph of the equation $y = x^2 + 4x - 12$?

- **A** $(-4, -18)$
- **B** $(-2, -16)$
- **C** $(0, -12)$
- **D** $(12, -2)$
- **E** $(2, 0)$

**24.** *Multiple Choice* What are the solutions of $0 = x^2 - 5x - 50$?

- **A** $-25, 2$
- **B** $2, 25$
- **C** $-5, -10$
- **D** $5, -10$
- **E** $-5, 10$

**25.** *Multiple Choice* Use the quadratic formula to find a solution for $-2x^2 - 3x + 9 = 0$.

- **A** $3$
- **B** $6$
- **C** $-6$
- **D** $-\frac{3}{2}$
- **E** $\frac{3}{2}$

**26.** *Multiple Choice* Use the discriminant to determine the number of solutions of the equation $3x^2 - 2x + 8 = 0$.

- **A** $1$
- **B** $2$
- **C** $3$
- **D** Infinitely Many
- **E** None

**27.** *Multiple Choice* Which ordered pair is a solution of the inequality $y > 2x^2 - 9x - 18$?

- **A** $(-1, -9)$
- **B** $(-3, 17)$
- **C** $(2, 6)$
- **D** $(0, -22)$
- **E** $(-2, 4)$

**28.** *Multiple Choice* Which of the following is equal to $(x + 7)(x - 3)$?

- **A** $x^2 + 4x - 21$
- **B** $x^2 + 10x - 21$
- **C** $x^2 - 10x - 21$
- **D** $x^2 - 10x + 4$
- **E** $x^2 - 4x - 21$

---

*Quantitative Comparison* In Exercises 29 and 30, add the solutions of the equations. Choose the statement below that is true about the given numbers.

- **A** The value in column A is greater.
- **B** The value in column B is greater.
- **C** The values are equal.
- **D** The relationship cannot be determined from the information given.

| | Column A | Column B |
|---|---|---|
| **29.** | $x^2 - 8x - 24 = 20$ | $x^2 + 24 = -10x$ |
| **30.** | $6x^2 - 7x + 2 = 0$ | $9x^2 - 36 = 0$ |

**31.** *Multiple Choice* Which of the following is the complete factorization of $6x^3 + 39x^2 - 21x$?

- **A** $(3x^2 + 7)(2x - 3)$
- **B** $3x(x + 7)(2x - 1)$
- **C** $3x(2x + 7)(x - 1)$
- **D** $3x(2x + 1)(x - 7)$
- **E** $(3x + 3)(2x^2 - 7x)$

---

*Quantitative Comparison* In Exercises 32–34, choose the statement below that is true about the given values of $x$.

- **A** The value in column A is greater.
- **B** The value in column B is greater.
- **C** The two values are equal.
- **D** The relationship cannot be determined from the information given.

| | Column A | Column B |
|---|---|---|
| **32.** | $\dfrac{3x}{5} = \dfrac{17}{4}$ | $\dfrac{5}{9} = \dfrac{2x}{25}$ |
| **33.** | 15% of 120 = $x$ | 85% of 290 = $x$ |
| **34.** | $\dfrac{x + 2}{18} = \dfrac{x}{8}$ | $\dfrac{2x - 3}{15} = \dfrac{x + 6}{12}$ |

---

**Algebra 1, Concepts and Skills**
Standardized Test Practice Workbook

**35. *Multiple Choice*** The variables $x$ and $y$ vary inversely. When $x$ is 10, $y$ is 7. If $x$ is 5, then $y$ is ___?___.

Ⓐ 21 Ⓑ 3.5 Ⓒ 350

Ⓓ $-7$ Ⓔ 14

**36. *Multiple Choice*** Simplify the

expression $\dfrac{x}{x+1} + \dfrac{1}{x}$.

Ⓐ $\dfrac{1}{x}$ Ⓑ $\dfrac{1}{x+1}$

Ⓒ $\dfrac{x^2 + x + 1}{x(x+1)}$ Ⓓ $\dfrac{x+2}{x+1}$

Ⓔ $\dfrac{3x+1}{x(x+1)}$

**37. *Multiple Choice*** Divide $(14x^2 + 19x - 3)$ by $(7x - 1)$.

Ⓐ $2x - 3$ Ⓑ $2x + 3$

Ⓒ $3x + 2$ Ⓓ $3x - 2$

Ⓔ $98x^3 + 119x^2 - 40x + 3$

**38. *Multiple Choice*** What is the domain of the function $y = 6 - \sqrt{x - 3}$?

Ⓐ $x \geq 6$ Ⓑ $x \leq 6$

Ⓒ $x \geq -3$ Ⓓ $x \geq 3$

Ⓔ $x \leq 3$

**39. *Multiple Choice*** Simplify $\dfrac{16}{2 - \sqrt{3}}$.

Ⓐ $32 + 16\sqrt{3}$ Ⓑ $\dfrac{32 + 16\sqrt{3}}{11}$

Ⓒ $\dfrac{18\sqrt{3} + 3}{11}$ Ⓓ $32 - 16\sqrt{3}$

Ⓔ $18 + \sqrt{3}$

**40. *Multiple Choice*** What term should be added to $x^2 - \frac{2}{3}x$ so that the result is a perfect square trinomial?

Ⓐ $\frac{4}{9}$ Ⓑ $\frac{4}{3}$ Ⓒ $-\frac{1}{9}$

Ⓓ $\frac{1}{9}$ Ⓔ $\frac{1}{3}$

**41. *Multiple Choice*** What is the unknown length of the triangle?

Ⓐ 4

Ⓑ 8

Ⓒ 16

Ⓓ 5

Ⓔ 24

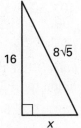

**42. *Multiple Choice*** What is the distance between $(7, -3)$ and $(-1, 4)$?

Ⓐ 6.1 Ⓑ 9.2 Ⓒ 8.1

Ⓓ 9.5 Ⓔ 10.6

**43. *Multiple Choice*** What is the length of side $a$ in the triangle?

Ⓐ 45

Ⓑ 6

Ⓒ 7

Ⓓ 8

Ⓔ 9

**Algebra 1, Concepts and Skills**
Standardized Test Practice Workbook

Chapter 12

# ANSWERS

## Chapter 1

**Lesson 1.1** **1.** D **2.** C **3.** A **4.** E **5.** E
**6.** C **7.** A **8.** B **9.** B **10.** B **11.** C
**12.** A

**Lesson 1.2** **1.** E **2.** B **3.** B **4.** A **5.** E
**6.** D **7.** D **8.** C **9.** C **10. a.** $12 \text{ ft}^3$
**b.** 89.76 gallons **c.** 67.32 gallons **d.** Yes

**Lesson 1.3** **1.** E **2.** D **3.** C **4.** A **5.** E
**6.** B **7.** B **8.** B **9.** E **10.** A **11.** C
**12.** B

**Lesson 1.4** **1.** B **2.** B **3.** B **4.** A **5.** D
**6.** C **7.** E **8.** A **9.** E **10.** B
**11. a.** $20w \geq 500$ **b.** 500 represents the
amount you need and 20 represents the savings
per week. **c.** at least 25 weeks **d.** at least
34 weeks

**Lesson 1.5** **1.** D **2.** B **3.** B **4.** A **5.** E
**6.** C **7.** E **8.** C **9.** B

**Lesson 1.6** **1.** A **2.** B **3.** C **4.** E **5.** D
**6. a.**

**b.** 20 miles **c.** about 6.2 mi/h **d.** about 6.7 mi/h

**Lesson 1.7** **1.** A **2.** D **3.** C **4.** C **5.** E
**6.** B **7.** A **8.** C

**Lesson 1.8** **1.** A **2.** C **3.** C **4.** E **5.** B
**6. a.** $1.25x = p$

**b.**

| $x$ | 50 | 100 | 150 | 200 |
|------|-------|-----|--------|-----|
| $P$ ($) | 62.50 | 125 | 187.50 | 250 |

**c.** $1.25x - 75 = p$

## Chapter 2

**Lesson 2.1** **1.** B **2.** D **3.** E **4.** E **5.** A
**6.** C **7.** C **8.** A **9.** B **10.** A **11.** A

**Lesson 2.2** **1.** E **2.** C **3.** E **4.** E **5.** E
**6.** B **7.** A **8.** D **9.** C **10.** B **11.** C
**12.** C

**Lesson 2.3** **1.** C **2.** D **3.** C **4.** A **5.** C
**6.** D **7.** E **8.** D **9.** B **10.** B **11.** A
**12.** C

**Lesson 2.4** **1.** B **2.** C **3.** D **4.** B **5.** D
**6.** D **7.** A **8.** E **9.** B **10.** C
**11. a.** $-2, 11, -2, -11$ **b.** a decrease in the
number of birds spotted **c.** an increase in the
number of birds spotted

**Lesson 2.5** **1.** A **2.** D **3.** B **4.** A **5.** A
**6.** A **7.** C **8.** E **9.** C
**10. a.** $75 - 3h = m$ **b.** $39 **c.** 25 hours

**Lesson 2.6** **1.** B **2.** A **3.** C **4.** A **5.** E
**6.** B **7.** C **8. a.** $4x + 4(2.5), 4(x + 2.5)$
**b.** yes

**Lesson 2.7** **1.** C **2.** D **3.** A **4.** D **5.** D
**6.** C **7.** B **8.** B **9. a.** A **b.** 21.3 mi

**Lesson 2.8** **1.** B **2.** C **3.** A **4.** B **5.** D
**6.** A **7.** A **8.** B **9.** A **10.** B **11.** A
**12.** A **13.** A

## Chapter 3

**Lesson 3.1** **1.** B **2.** B **3.** C **4.** D **5.** A
**6.** C **7.** D **8.** C **9.** A **10.** B **11.** A
**12.** B **13.** D **14.** A

**Lesson 3.2** **1.** C **2.** D **3.** C **4.** A **5.** A
**6.** B **7.** E **8.** D **9.** A **10.** C **11.** A

**Lesson 3.3** **1.** B **2.** C **3.** A **4.** A **5.** D
**6.** E **7.** D **8.** B **9.** C **10.** C **11.** B
**12.** A **13.** C

**Lesson 3.4** **1.** C **2.** B **3.** A **4.** E **5.** D
**6.** B **7.** C **8.** D **9.** D **10.** B **11.** C

*Lesson 3.5* **1.** D **2.** A **3.** E **4.** B **5.** B
**6.** B **7. a.** C **b.** 60 **c.** You must use the pool more than 60 days to justify the cost of becoming a member.

*Lesson 3.6* **1.** C **2.** E **3.** E **4.** D **5.** B
**6.** A **7.** B **8.** B **9.** C
**10. a.** B **b.** 82 **c.** 52

*Lesson 3.7* **1.** B **2.** D **3.** C **4.** C
**5.** E **6.** D **7.** B **8.** C **9.** A

*Lesson 3.8* **1.** D **2.** B **3.** B **4.** C **5.** D
**6.** D **7.** A **8.** C **9. a.** Divide the total cost by the total number of gallons of gas. **b.** $1.51
**c.** $24.16

*Lesson 3.9* **1.** B **2.** D **3.** A **4.** E **5.** B
**6.** C **7.** E **8.** E **9.** C **10.** A **11.** C
**12.** B **13.** C

## Chapter 4
*Lesson 4.1* **1.** B **2.** A **3.** E **4.** C **5.** B
**6.** C
**7.**
**a.**

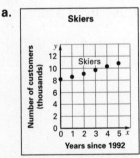
Skiers

**b.** The number of skiers is increasing over time.

*Lesson 4.2* **1.** D **2.** B **3.** C **4.** A **5.** C
**6.** C **7.** A **8.** C **9. a.** $y = -2x + 36$

**b.**

| $x$ | 5 | 10 | 15 |
|-----|---|----|----|
| $y$ | 26 | 16 | 6 |

**c.**
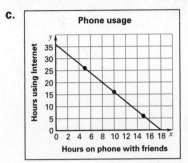

Phone usage

Hours using Internet

Hours on phone with friends

*Lesson 4.3* **1.** C **2.** E **3.** B **4.** D **5.** E
**6.** A **7.** A **8. a.** The function has a constant value, 3, regardless of the value of $x$. **b.** all real numbers **c.** 3 **d.** $(-4, 3)$

*Lesson 4.4* **1.** B **2.** E **3.** E **4.** C **5.** B
**6.** E **7.** D **8.** B **9.** B **10.** A

*Lesson 4.5* **1.** C **2.** D **3.** A **4.** B **5.** A
**6.** E **7.** C **8.** B **9.** C **10.** B **11.** A
**12.** D

*Lesson 4.6* **1.** A **2.** B **3.** D **4.** E **5.** D
**6.** C **7.** C **8.** B **9.** A **10.** C

*Lesson 4.7* **1.** C **2.** B **3.** C **4.** D **5.** C
**6.** D **7.** C **8.** E **9.** B **10.** C **11.** A

*Lesson 4.8* **1.** C **2.** A **3.** D **4.** D **5.** E
**6.** B **7.** C **8.** B **9.** A **10.** A

## Chapter 5
*Lesson 5.1* **1.** B **2.** E **3.** A **4.** C **5.** B
**6.** B **7.** B **8.** C **9.** A

*Lesson 5.2* **1.** D **2.** B **3.** A **4.** B **5.** E
**6.** C **7.** B **8.** A **9.** B

*Lesson 5.3* **1.** D **2.** A **3.** D **4.** E **5.** B
**6.** E **7.** A **8.** A **9.** A

*Lesson 5.4* **1.** D **2.** E **3.** B **4.** C **5.** A
**6.** C **7.** B **8.** A **9.** B **10.** A

*Lesson 5.5* **1.** B **2.** E **3.** C **4.** C **5.** A
**6.** A **7. a.** B **b.** 8, 5, 2 **c.** 11:20 A.M.

# Chapter 5 *continued*

**Lesson 5.6** **1.** B **2.** D **3.** C **4.** E **5.** B
**6.** C **7.** B **8. a.** $\overline{AB}: y = \frac{5}{2}x + \frac{11}{2}$
$\overline{BC}: y = -\frac{2}{5}x + \frac{13}{5}$; $\overline{AC}: y = \frac{3}{7}x - \frac{5}{7}$
**b.** $\overline{AB}$ and $\overline{BC}$; perpendicular slopes

# Chapter 6

**Lesson 6.1** **1.** A **2.** C **3.** C **4.** C
**5.** D **6.** E **7.** D **8. a.** $b \geq 250$ **b.** No
**c.** *Sample answer:* 250, 300, 350

**Lesson 6.2** **1.** B **2.** E **3.** D **4.** C **5.** B
**6.** A **7.** B **8.** A **9.** D

**Lesson 6.3** **1.** B **2.** A **3.** A **4.** D **5.** C
**6.** C **7. a.** $x \geq 660$ **b.** $x \geq 165$ **c.** $x \geq 7$

**Lesson 6.4** **1.** A **2.** C **3.** C **4.** D **5.** B
**6.** E **7.** B **8.** B **9.** D

**Lesson 6.5** **1.** B **2.** D **3.** A **4.** D **5.** C
**6.** B **7. a.** 48, 32, 16, 0, $-16$, $-32$ **b.** The
velocity decreases until the ball reaches its highest
point at $t = 1.5$. The ball then falls downward and
its velocity increases. **c.** $t < 0.5$ or $t > 2.5$

**Lesson 6.6** **1.** B **2.** D **3.** D **4.** E **5.** A
**6.** B **7.** D **8.** B **9.** A

**Lesson 6.7** **1.** D **2.** A **3.** A **4.** A **5.** E
**6.** C **7.** B **8.** A **9.** C

**Lesson 6.8** **1.** D **2.** A **3.** B **4.** C **5.** B
**6.** E **7. a.** $8.5x + 10.5y \leq 485$
**b.**

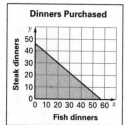
Dinners Purchased

**c.** 30 or less
**d.** No, you can't order
fractional parts of a
meal.

# Cumulative Review

**Chapters 1–6** **1.** B **2.** C **3.** A **4.** E **5.** B
**6.** A **7.** A **8.** D **9.** C **10.** E **11.** E

**12.** C **13.** B **14.** C **15.** B **16.** A **17.** A
**18.** E **19. a.** $32 + 12(x - 1)$ **b.** 3
**20.** D **21.** E **22.** A **23.** B **24.** D **25.** D
**26.** B **27.** C **28.** D **29.** E **30.** B **31.** A
**32.** C **33.** A **34.** A **35.** D **36.** B **37.** B
**38.**
**a.**
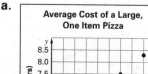
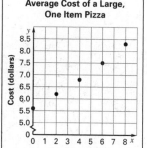

**b.** The cosst of a
large one item
pizza has
increased
steadily from
1990 to 1998.

**39.** B **40.** C **41.** B **42.** A **43.** C
**44. a.** 6 **b.** $y = 6x + 28$ **c.** 100

# Chapter 7
**Lesson 7.1** **1.** E **2.** C **3.** B **4.** C **5.** D
**6.** D **7.** B **8.** A **9.** B **10.** A

**Lesson 7.2** **1.** B **2.** C **3.** E **4.** B **5.** A
**6.** D **7.** C **8.** A **9.** C **10.** B

**Lesson 7.3** **1.** B **2.** D **3.** E **4.** C **5.** A
**6. a.** $3x + 2y = 1.70$ **b.** $(0.40, 0.25)$ **c.** \$.25
$2x + 3y = 1.55$

**Lesson 7.4** **1.** B **2.** C **3.** C **4.** E **5.** D
**6.** D **7.** A **8.** C **9.** B

**Lesson 7.5** **1.** A **2.** B **3.** A **4.** B **5.** D
**6.** B **7.** A **8.** C **9.** B

**Lesson 7.6** **1.** A **2.** D **3.** B **4.** D **5.** E
**6. a.** $x + y \leq 25$ **b.**
$5x + 8y \geq 150$
$x \geq 0$
$y \geq 0$

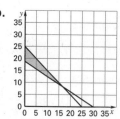

**c.** Answers may vary.
*Sample answer:* 15
hours at first job and 10 hours at second job, or
5 hours at first job and 18 hours at second job.

## Chapter 8

**Lesson 8.1** **1.** B **2.** C **3.** D **4.** E **5.** A
**6.** B **7.** C **8.** D **9.** B **10.** C **11.** B
**12.** D

**Lesson 8.2** **1.** C **2.** B **3.** B **4.** A **5.** E
**6.** B **7.** D **8.** C **9.** A **10.** C **11.** A
**12.** B

**Lesson 8.3** **1.** A **2.** D **3.** C **4.** A **5.** D
**6.** D **7. a.** $367.51 **b.** $500 **c.** $583.20

**Lesson 8.4** **1.** B **2.** E **3.** C **4.** D **5.** A
**6.** A **7.** D **8.** D **9.** E **10.** D **11.** A
**12.** C **13.** A

**Lesson 8.5** **1.** A **2.** D **3.** B **4.** D **5.** E
**6.** C **7.** B **8.** C **9.** A **10.** A **11.** B
**12.** B

**Lesson 8.6** **1.** C **2.** D **3.** B **4.** B **5.** C
**6.** D **7. a.** $y = 50,000(1.15)^t$ **b.** $133,001
**c.** $37,807 **d.** Answers may vary. *Sample
answer:* It will probably continue to increase, but
not as rapidly.

**Lesson 8.7** **1.** E **2.** C **3.** C **4.** A **5.** C
**6.** D **7.** A **8.** C **9.** A

## Chapter 9

**Lesson 9.1** **1.** D **2.** E **3.** D **4.** D **5.** E
**6.** A **7.** B **8.** E **9.** E **10.** A **11.** C
**12.** A

**Lesson 9.2** **1.** B **2.** D **3.** C **4.** D **5.** C
**6.** A **7.** C **8.** B **9.** B **10.** B

**Lesson 9.3** **1.** D **2.** C **3.** E **4.** A **5.** D
**6.** B **7.** D **8.** A **9.** B **10.** C **11.** A

**Lesson 9.4** **1.** D **2.** B **3.** B **4.** C **5.** E
**6.** C **7.** D **8.** A **9.** A **10.** C **11.** B

**Lesson 9.5** **1.** D **2.** A **3.** C **4.** B **5.** B
**6.** A **7.** C **8.** A **9.** C **10.** E **11.** D

**Lesson 9.6** **1.** B **2.** E **3.** B **4.** C **5.** B
**6.** A **7.** E **8.** B
**9. a.** $h = -16t^2 + 50$ **b.** about 1.77 sec
**c.** *Sample answer:* The speed of the ski lift if it is
moving when the keys are dropped, the angle at
which the keys are dropped, and the wind and
weather conditions.

**Lesson 9.7** **1.** B **2.** E **3.** C **4.** C **5.** B
**6.** C **7.** C **8.** A **9.** C **10.** B

**Lesson 9.8** **1.** B **2.** E **3.** A **4.** D **5.** A
**6. a.** 4 feet **b.** 4 feet **c.** Not if the water line
is outside of $y \leq x^2 - 2x - 3$.

## Chapter 10

**Lesson 10.1** **1.** E **2.** A **3.** D **4.** B **5.** B
**6.** B **7. a.** $S = 8.9t^2 + 1.2t + 73$
**b.** $111,000 **c.** $350 = 21.1t^2 + 7.5t + 138$

**Lesson 10.2** **1.** D **2.** E **3.** A **4.** E **5.** D
**6.** C **7.** A **8.** B **9.** B **10.** A

**Lesson 10.3** **1.** E **2.** B **3.** D **4.** E **5.** E
**6.** B **7.** A **8.** A **9.** B

**Lesson 10.4** **1.** E **2.** B **3.** A **4.** C **5.** C
**6.** D **7.** C **8. a.** 12 feet **b.** 6.48 feet
**c.** 6 people

**Lesson 10.5** **1.** B **2.** C **3.** E **4.** C **5.** A
**6.** D **7.** D **8.** C **9.** A **10.** A **11.** C
**12.** A

**Lesson 10.6** **1.** B **2.** D **3.** C **4.** B **5.** A
**6.** C **7.** E **8.** D
**9. a.** $0 = -16t^2 + 4t + 20$ **b.** $\frac{5}{4}$ and $-1$
**c.** The correct answer is 1.25 seconds. Time
cannot be negative for this problem.

**Lesson 10.7** **1.** D **2.** A **3.** B **4.** C **5.** D
**6.** D **7.** A **8.** E **9.** B **10.** C **11.** B

**Lesson 10.8** **1.** E **2.** B **3.** A **4.** B **5.** C
**6.** D **7. a.** 2 seconds **b.** about 2.8 seconds

# Chapter 11

## Chapter 11
**Lesson 11.1** **1.** E **2.** C **3.** A **4.** D **5.** B
**6.** C **7.** A **8.** B **9.** A **10.** D

**Lesson 11.2** **1.** E **2.** A **3.** D **4.** D **5.** D
**6.** B **7.** C

**8. a.** Sample table:

| $h$ | 240 | 120 | 100 |
|---|---|---|---|
| $r$ | 5.00 | 10.00 | 12.00 |

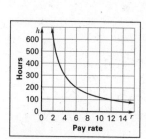

**b.** inverse; $y = \dfrac{1200}{x}$

**c.** The model is similar: $y = \dfrac{800}{x}$.

**Lesson 11.3** **1.** D **2.** B **3.** E **4.** D **5.** E
**6.** C **7.** B **8.** A **9.** B **10.** C

**Lesson 11.4** **1.** E **2.** B **3.** A **4.** B **5.** D
**6.** C **7.** A **8.** B **9.** C

**Lesson 11.5** **1.** D **2.** E **3.** B **4.** D **5.** C
**6.** B **7.** A **8.** A

**Lesson 11.6** **1.** E **2.** E **3.** D **4.** B **5.** A

**6. a.** $\dfrac{1200}{(x+25)} + \dfrac{1200}{(x-30)}$

**b.** $\dfrac{2400x - 6000}{(x+25)(x-30)}$ **c.** about 12.4 hours

**d.** The smaller plane is traveling at half the speed of the larger plane. At first glance, the trip might seem to be exactly twice as long, but due to wind force the trip actually takes about 26.7 hours.

**Lesson 11.7** **1.** B **2.** B **3.** A **4.** C **5.** E
**6.** E **7.** B **8.** A **9.** A

## Chapter 12
**Lesson 12.1** **1.** A **2.** A **3.** B **4.** A **5.** C
**6.** C **7.** C **8.** B **9.** B

**Lesson 12.2** **1.** D **2.** D **3.** E **4.** B **5.** C
**6.** E **7.** B **8.** B **9.** C

**Lesson 12.3** **1.** D **2.** C **3.** E **4.** A **5.** B
**6.** A **7.** C **8.** D **9.** B **10.** C **11.** A
**12.** B

**Lesson 12.4** **1.** C **2.** C **3.** D **4.** A **5.** B
**6.** E **7.** B **8.** D **9.** A **10.** B **11.** A

**Lesson 12.5** **1.** D **2.** C **3.** B **4.** B **5.** C
**6.** B **7.** D **8.** E **9.** C **10.** A **11.** B

**Lesson 12.6** **1.** C **2.** A **3.** B **4.** A **5.** D
**6.** E **7.** D **8.** D **9. a.** $d + 2$

**b.**

**c.** Kim, 6 mi; Cindy, 8 mi **d.** *Sample answer:* I used factoring because the resulting quadratic equation could be factored.

**Lesson 12.7** **1.** D **2.** B **3.** A **4.** C
**5.** C **6.** D **7.** D **8.** A **9.** C **10.** A

**Lesson 12.8** **1.** E **2.** B **3.** A **4.** B **5.** C
**6.** D **7. a.** $\left(-\frac{1}{2}, -1\right)$ **b.** $\frac{13}{2}$ **c.** $\frac{13}{2}$
**d.** 13; They are equal.

**Lesson 12.9** **1.** D **2.** D **3.** C **4.** E **5.** A
**6.** B **7.** D **8.** A
**9.** Sample answers are given.

   **a.** $(6 - 4) - 1 = 1$, but $6 - (4 - 1) = 3$.
   **b.** $1^2 = 1$ **c.** $a = 7, b = 5, c = 0$, **d.** $a = 3$

## Cummulative Review
*Chapters 1–12* **1.** B **2.** E **3.** B **4.** A
**5.** A **6.** C **7.** D **8.** B **9.** D **10.** E
**11.** E **12.** B **13.** C **14.** D **15.** A **16.** E
**17.** B **18.** B **19.** A **20.** D **21.** C **22.** C
**23.** B **24.** E **25.** E **26.** E **27.** C **28.** A
**29.** A **30.** A **31.** B **32.** A **33.** B **34.** B
**35.** E **36.** C **37.** B **38.** D **39.** A **40.** D
**41.** B **42.** E **43.** C